THROUGH EMOTIONS TO MATURITY

Originally published in 1982 as *Wege aus Angst und Symbiose:*
Marchen psychologisch gedeutet
Copyright © 1982 Walter-Verlag AG, Olten, Switzerland

Manufactured in the United States of America
Printed on acid-free, recycled paper

First U.S. Edition

LIBRARY OF CONGRESS CATALOGING-IN-PUBLICATION DATA
Kast, Verena, 1943–
[Wege aus Angst und Symbiose. English]
Through emotions to maturity: psychological readings of fairy tales / Verena Kast.
p. cm.
Includes bibliographical references.
ISBN 0-88064-205-X (cloth: alk.paper): $19.95.
ISBN 0-88064-206-8 (pbk.: alk. paper): $11.95.
1. Fairy tales—History and criticism. 2. Psychoanalysis and folklore.
3. Anxiety. 4. Symbiosis (Psychology) I. Title.
GR550.K38413 1993
398.21—dc20   93-33193 CIP

THROUGH EMOTIONS TO MATURITY
Psychological Readings of Fairy Tales

VERENA KAST

Translated by Douglas Whitcher
with Susan C. Roberts

FROMM INTERNATIONAL
PUBLISHING CORPORATION

CONTENTS

FOREWORD TO THE ENGLISH EDITION   i

FOREWORD TO THE GERMAN EDITION   vii

GETTING THROUGH ANXIETY

INTRODUCTION   1

THE LAD WHO FEARED NOTHING   2
To Be Human Is to Have Fear

THE GOOSE MAID   24
Fear in Separating from the Mother

GRAYCOAT   46
Fear in Separating from the Father

NIXIE IN THE POND   63
Fear of Overwhelming Emotions

GETTING THROUGH SYMBIOSIS

INTRODUCTION   85

JOURNEY TO THE UNDERWORLD   100
THROUGH THE HELLISH WHIRLPOOL OF FAFÁ
Contending with the Devouring Primal Ground

REDHAIR GREENEYES   120
A Way out of Father-boundedness

THE DAUGHTER OF THE LEMON TREE   140
A Way out of Overprotection

JORINDA AND JORINGEL   160
A Way out of Infatuation into Relationship

CONCLUDING REMARKS   168

NOTES   173

BIBLIOGRAPHY   175

■ Foreword to the English Edition

BY DOUGLAS WHITCHER

This book is another stepping stone for English-speaking readers toward greater familiarity with the thought of Verena Kast. Until now we have become acquainted with Kast as a psychologist of the mourning process (1988), a mythographer of relationship (1986), a theorist and practitioner of depth crisis intervention (1990), a Dionysian philosopher of joy (1991), a Sisyphusian mid-life counselor (1991), an educator who is able and willing to readably summarize the state of the art of contemporary Jungian psychotherapy (1992), and an initiator into the world of Active Imagination (1993). Recently, a publication appeared together with the authors Ingrid Riedel and Mario Jacoby in which we were offered a sample of Kast as an interpreter of folktales (1992). The present volume is the first to appear in English that is devoted exclusively to her reflections on folktales. Fromm International will bring us in time all six of Kast's books on folktales. If her publishing history in German is any indication, we may predict that Verena Kast's thought will soon become well known to her English-speaking audience as well.

Kast's popularity has been held against her. How can anything be profound that is popular? And yet Jungian psychology itself provides the answer to this apparent riddle: If archetypal images express dilemmas that are common to most if not all of humanity, and if folktales are an expressive medium through which these images are entertainingly passed along, then it should come as no surprise that a therapist who writes a great deal about folktales would sell a great many books, provided she can write in a way that does not bore her audience.

Of course not everyone can write about folktales without boring their audiences. To my mind, one of Kast's unique gifts is her ability to spring deftly from folktale to therapy and back again. Sometimes one hardly notices the transition because her parallels are anything but contrived. We do not read about dreams reproducing secret chapters from the history of the human spirit; we hear how a folktale image motivated a client to change.

I was introduced to the present book by an analysand in Zurich who was caught in a symbiotic tie with his mother. He was her pride and glory—when he succeeded in the things she thought were most important—and her shame and undoing when he failed. At the time, he was failing in a big way—as far as she was concerned. Now that I have actually read the book myself, I realize that by telling me about this book he was trying to tell me that his failings in one department might yet lead to development in another. Perhaps he would have to subject her to an undoing in order to do something for himself. From Kast's perspective, undoing himself from her would be doing her a favor as well.

I suspect that the analysis would have been helped along if I had actually read Kast's book. Perhaps then we could have talked about the boy in "The Daughter of the Lemon Tree" who was so spoiled that there was nothing he seemed to be able to accomplish on his own, about how it was only through his careless games and passive aggression that he smashed an old lady's cooking pot, and that it was only by means of his old lady's frustration and curse that he was able to discover his own steadfast will. Kast may not resemble such an old lady by her outward appearance, but her advice to the young man reminds me of what I could imagine her telling a young man in therapy whose dreams are not bothered much by reality: He needs to trudge about in iron shoes for no less than three years. This may sound like an odd and impractical sentence, but it may be the only way to ground someone to the soil from which his personal identity will eventually grow, if only he can stick to it long enough.

Or we could have talked about the young man in "Redhair Greeneyes" who went ahead and did the one thing his tired-out old father told him not to do. The young man in analysis was still desperately trying to please a father that he imagined for himself out of the whole cloth of his disappointments. He believed that his father's alcoholic ruin translated into the advice, "Whatever you do, don't do what I did." And yet by trying to avoid the various apples of temptation that presented themselves along his walk through life, he walked right into the same old traps of self-destruction—unconsciously, unlike the protagonists of the folktales that Dr. Kast has interpreted for us here. Rather than avoiding the demon that had ruined his father's life, he would have done well to "hire" his father's devil, like the young man in the folktale. Then he would have taken the path of risk, the only path that could lead him through the perils that had stopped his life dead in its tracks. He could have applied a bit of poison to his mindset with the aim of killing off his self-castrating expectations.

It is of course false to say that the young merchant "hired" his father's devil, for this devil told the youth what to do at every turn. And yet, neither would it be correct to say that the devil took possession of the young man. At the very end of the tale, we learn that this demon was actually a shade the young man had ransomed, the shadow that his father had never allowed to enter his own life. The possibility of redeeming one's father's shadow is good news for young men in analysis these days.

"Good news" indeed. There are those who would object to Kast's optimistic reading of psychology through the lens of folktales; it reminds them of some "happy ending" Gospel. What does this gospel have to do with Jung's vision of a wrenching clash of opposites that never finds an end but only winds its way towards new heights of tragic conflict, they ask. Is Kast suggesting that problems can actually be solved, that folktales can provide us with maps for finding our way through emotions to maturity, as the title suggests?

Kast's "belief," however outlandish or optimistic, is nevertheless critical and psychological, namely, that relationships become diseased, stunted in their growth, packed full of unrealistic expectations, and that it is the job of the psychotherapist schooled in folklore to unravel the eerie images that implicate themselves in such relationships. The only way out of the image is through the image, just as the only way out of symbiosis is through symbiosis, the only way out of anxiety through anxiety.

This may sound like a recipe—the client, like the protagonist, has only to endure in order to prevail in the end over evil and injustice. Therapists, like folktales, may seem to preach simplistic convictions, like the benefit of following one's dreams, as in "The Nixie in the Pond," or the rewards of being patient, as in "The Goose Maid." If we adopt a folkloric, therapeutic view of life, don't we run the risk of ruining our lives by pinning our hopes on rhetorically persuasive illusions?

Yet to unravel their seductive power over the soul, as Joringel does when he frees Jorinda from her entrapment as a nightingale, it is not enough to decry the seductive allure of images. One must know the ways of this "witch" very intimately. Kast can be of great assistance in this task. Although Joringel's fate and official task was to tend sheep, his secret pastime was a study of the witch that held his love captive. Then one day he was bold enough to crash the gates of his romantic desire and recover a real woman from the trap of his happily-ever-after symbiosis. Through Kast's interpretations, the poison of images and illusions—indeed of "fairy tales"—becomes its own antidote.

Kast, Verena, *The Nature of Loving: Patterns of Human Relationship*, translated by Boris Matthews (Wilmette, Illinois: Chiron, 1986).

_____ , *A Time to Mourn: Growing through the Grief Process*, translated by Diana Dachler & Fiona Cairns (Einsiedeln, Switzerland: Daimon, 1988).

_____ , *The Creative Leap: Psychological Transformation Through Crisis*, translated by Douglas Whitcher (Wilmette, Illinois: Chiron, 1990).

_____ , *Joy, Inspiration, and Hope*, translated by Douglas Whitcher (College Station: Texas A & M University, 1991).

_____ , *Sisyphus: The Old Stone—A New Way*, translated by Norman M. Brown (Einsiedeln, Switzerland: Daimon, 1991).

_____ , *The Dynamics of Symbols: Fundamentals of Jungian Psychotherapy*, translated by Susan A. Schwarz (New York: Fromm International, 1992).

_____ , Jacoby, Mario and Riedel, Ingrid *Witches, Ogres, and the Devil's Daughter: Encounters with Evil in Fairy Tales*, translated by Michael H. Kohn (Boston and London: Shambhala, 1992).

_____ , *Imagination as Space of Freedom: Dialogue between the Ego and the Unconscious*, translated by Anselm Hollo (New York: Fromm International, 1993).

■ Foreword to the German Edition

BY VERENA KAST

The essays in this volume were first presented as lectures at the psychotherapy conferences in Lindau, Switzerland, in 1980 and 1981. The first series concerned ways out of symbiosis and the second, anxiety and ways to deal with it in folktales. On the recommendation of my Swiss publisher, I decided to bring the lectures from both series together in one volume. Thus, in my search for a title, I felt it made sense to compare symbiosis with anxiety, and so I ended up with, "Ways out of Anxiety and Symbiosis."* The comparison was intended to point to a relationship between the two concepts that could be summed up as follows: Typically, a person seeks symbiosis as a way of avoiding anxiety—especially the fear of separation. Thus, ways out of symbiosis are always ways of coping with and of getting through anxiety.

But why use folktales as the means to explore ways out of symbiosis and anxiety? In Jungian psychology, we view folktales as symbolic representations of problems common to most humans, as well as portrayals of possible ways to solve those problems. Folktales deal with something that breaks down the onward flow of life—a problem usually described in the tale's initial situation—and suggest a developmental path through and beyond the problem toward a new life. We all know that in the course of maturing, we make detours, expose ourselves to risks, and get completely tripped up. Our paths through life are no less dangerous than those encountered by the characters in

---

* The literal translation of the original title. (Editor's note.)

folktales. We of the Jungian school view the protagonists of folktales as role models, demonstrating for us how we might endure problematic situations or persist along a path on which possibilities for solving problems inevitably arise. In approaching such tales, we employ a "subjective level" of interpretation that we have adapted from our work with dreams, in which every figure that appears can be viewed as a personality trait of the dreamer. By the same token, each character in the folktale can be seen as a personality trait of the protagonist. For example when a woman meets a witch in a folktale, we say that she is encountering her own witchy side.

In our interpretations, we try to enter the folktale's narrative structure, its overall line of development, not simply those scenes in which the protagonist acts or is acted upon. We also give serious consideration to symbols, employing the method of amplification to discover what a given symbol might mean. What this means is that we attempt to place the symbol in the context of parallel tales that fall under the same motif-rubric, we reflect on moments in history in which the symbol performed some function, and consider the interpretive contexts in which it has been used. By means of amplification, the general outline of a symbol's meaning emerges .[1]

Folktales have been interpreted in many different ways, according to the methods of such disciplines as the study of German languages and literature, the study of folklore, sociology, and psychology. The folktale can be approached from all of these perspectives. Different aspects of the folktale assume greater importance, depending on the angle taken. It would of course be ideal if we could take all viewpoints into consideration, as well as include the thoughts of a number of commentators, since the interpreter's own personality has a large influence on the interpretation. Nor can we afford to overlook the fact that folktales, since they are composed of images, are never one-dimensional. As more and more levels emerge, the image becomes increasingly mysterious, and it becomes harder and harder to settle on one definitive interpretation. This is what makes the interpretation of folktales so exciting; the possibilities are never exhausted. The criterion I use for a successful and defensible interpretation is whether it is consistent and effective in giving meaning to a number of different details that would not otherwise make sense. At the very least, a given interpretation should stimulate reflection or uncover contradictions. In any case, there is no such thing as a "correct" interpretation.

Interpreting folktales is neither the sole nor the most important way of encountering and engaging with them. Using active imagination, painting, drawing, or meditating on them are equally important methods for letting folktales stimulate emotional growth.

Getting Through Anxiety

■ Introduction

Fairy tales often deal with anxiety without speaking about it directly; in fact anxiety is seldom mentioned in such tales. But when we listen to a tale and immerse ourselves in its imagery, we often find ourselves fearing for the hero or heroine: there is the threat of poisoning that awaits Snow White at every turn, the abandonment suffered by Hansel and Gretel, the torrent of rage the witch hurls at Gretel. Once the protagonist survives the danger, we breathe a sigh of relief, for we too have survived another anxiety attack. From this perspective, then, there is hardly a fairy tale that does not deal with anxiety. Looking at ourselves through the fairy tale—which presents us with typical human dilemmas and allows us to imagine paths out of them—we see that we are confronted with anxiety every step of the way.

Anxiety belongs to the human condition; if we embrace the logic of all those tales in which someone "sets out to learn fear" we might even say that it is what makes us human. The first tale in this collection of interpretations—an Icelandic tale called "The Lad Who Feared Nothing"— provides the clearest example of this "education in anxiety." However, I also will be exploring the theme in essays on the Grimm Brothers' stories "The Goose Maid" and "The Nixie of the Pond" and the German folktale "Gray Coat."

■ The Lad Who Feared Nothing

TO BE HUMAN IS TO HAVE FEAR

There was once a very cheeky lad who was not afraid of anything. This was of great concern to his parents and other relatives, for no matter what they asked him to do, he never had the slightest fear that something bad might happen to him. At last they gave up and brought him to the village pastor, who they had decided should be the one to teach him fear.

But once the fellow was taken in, it soon became clear that he was not about to learn fear here either, try though the pastor might. The boy was no more rebellious or disrespectful with the pastor than he had been with his relatives. But still the time passed and the pastor's efforts to teach the lad fear proved to be in vain.

One winter's day, however, the pastor finally saw his opportunity to teach the boy some fear. As it happened, there were three corpses at the church awaiting burial. Since the bodies had been delivered late in the evening, they had to be stored in the sanctuary overnight. In those days, it was still proper to bury a corpse without a coffin, and so these were simply wrapped in shrouds. The pastor had them dragged into the church and then left them stretched out across the aisle, one beside the other, with very little space in between. That evening in the parsonage, the pastor told the boy, "Go quickly to the church, my son, and get me the book that is lying on the altar."

The boy—who was obliging even if he was brash—did what he was told right away. He went to the church, unlocked the door, and began making his way up the aisle. But after a few steps he tripped over something. After feeling around a bit, he realized it was a human body. Unfazed, he promptly shoved the body between the pews in order to get it out of the way. Then he continued on up the aisle, tripping over the second corpse, which he dealt with in the same nonchalant way as he had the first. The same thing happened with the third corpse, which he likewise shoved out of the aisle into the space between the pews. When he finally reached the altar, he took the book the pastor had wanted, then merrily walked back down the now-clear aisle, locked the church again, and returned to the parsonage. When the pastor asked if he had noticed anything unusual, the boy replied that he had not.

"You mean you didn't notice the corpses lying in the aisle? I forgot to tell you about them," the pastor said.

"Ah—the corpses," the boy replied. "Yes, I saw them. I didn't know what you were talking about at first, Father."

"Well, and how did you notice them?" asked the pastor. "Weren't they in your way?"

"Oh, that was no problem," the boy said.

"So how did you get past them?"

"I pushed them out of the aisle and into the spaces between the pews, which is where they are now."

At that, the pastor refrained from asking questions and just shook his head. The next morning he told the lad, "You must go away from here. I can't have you in my house any longer. You are godless and have no respect for the peace of the dead." The boy had nothing to say to this and so bid the pastor and all the townsfolk a polite farewell.

Now he wandered about for some time, not knowing where to stay. Stopping by an inn one night, he heard that the bishop of Skalholt had died. So he made a small detour there and, arriving in the evening, requested lodging at the bishop's house. The people there were glad to give him room and board but warned that he would have to look out for his own safety. What was there to be afraid of? the boy asked. Since the death of the bishop, things in Skalholt had taken a nasty turn, they explained. When nightfall came, ghosts appeared, making it impossible for anyone to remain on the premises. "All the more reason to stay," the boy answered.

The people urged him not to talk so recklessly and cautioned him that it really would not be fun to be around when the haunting began. As it grew dark, everyone left except the lad. With heavy hearts, they bid him farewell, believing that they would never see him again.

The young fellow stayed on alone and enjoyed himself. When it grew dark, he lit a lantern and wandered through the entire house, inspecting it. At last, he came into the kitchen. How well stocked it was! Fat lamb carcasses hung in a row, and there were many other delicious provisions as well. Not having eaten dried meat for some time, the boy developed quite an appetite for it from looking at the great quantities hanging in the kitchen. So he split some wood, made a fire, put a pot on to boil, and carved up a sheep carcass.

So far, there was no sign of ghosts. But once the meat was in the pot, the boy heard a muffled voice coming from the chimney above. "May I drop?" it asked.

"Why shouldn't you drop?" replied the boy. At this, the upper third of a man's body fell down the chimney—a head and shoulders, complete with arms and hands. It remained on the floor without moving for a time. Then the boy once again heard a deep voice echoing through the chimney. "May I drop?" it asked.

"Why shouldn't you drop?" the lad responded again. This time the trunk of a man fell out of the chimney, coming to rest next to the first piece.

Yet a third time the young man heard the question, "May I drop?" And once again he was cordial in his invitation: "Why shouldn't you drop? Perhaps you need something to stand on!" No sooner had he said this than down came the legs of a man.

They were incredibly large, of a size to match the other parts. Now all the pieces lay there on the floor together quietly. This was boring for the boy, so he began to step on them, asking them, "Now that you are all together, why don't you take a walk?" At this, the pieces assembled themselves into a gigantic man who walked out of the kitchen without saying a word.

The boy followed the huge man into the front room. There, he watched as the man unlocked a large chest full of coins. The ghost took up one handful after another, then threw the coins over his shoulder onto the floor. He kept doing this until the chest was empty. Then he reached into the pile that had built up behind him and began throwing the coins back over his shoulders into the chest. The boy stood by for the duration of the strange game, watching the gold coins rolling about on the floor.

As the night drew to a close, the ghost picked up his pace. Now he was frantically throwing money back into the chest and hurriedly sweeping up any coins that had rolled away. Soon the boy realized that the ghost was rushing because he knew dawn was approaching. And indeed, once all the money was back in the chest, the ghost made great haste to leave. But the boy stopped him, assuring him that there was no need to rush. "Yes there is," said the ghost. "It is almost daybreak." The ghost now tried to get around the boy, but the boy blocked the way. The ghost got angry and grabbed the lad. Bad things would happen if the boy got in his way, the ghost threatened. It then began fiercely beating up on the boy. Realizing that he was by far the weaker of the two, the boy did his best to avoid the heaviest blows and stay on his feet.

Things continued in this way for some time. Finally, with his back to the open door, the ghost attempted to lift the lad up in order to throw him down to the floor with one terrible crash. Seeing that his death was imminent, the boy resorted to cunning. With all his might, he threw himself against the ghost, knocking it backwards into the open doorway. The moment the ghost's head landed outside, daylight shone in his eyes, and he split into two and disintegrated.

Though a bit stiff and bruised, the boy went right to work making two wooden crosses, which he stuck into the ground on either side of the threshold to mark where the ghost's two halves had disappeared. Then he lay down and slept until the bishop's folk returned.

When they saw that the lad was still alive, they greeted him enthusiastically and asked him if he hadn't seen any ghosts in the night. He said he hadn't noticed any, but they refused to believe him.

That day, he stayed on at the bishop's house and rested. Not only was he too exhausted to move after his fight with the ghost, the bishop's folk wouldn't let him go, for he had restored their courage. When they got ready to leave for the night, the boy tried to assure them that the ghost wouldn't harm them anymore. But it was no use—they went their way. At least they didn't worry about him as much this time. Indeed, that night the boy slept soundly, and when the folk returned the next morning and inquired about the ghost, he assured them that he hadn't noticed anything. Then he told them all about what had happened on the first night and showed them the wooden crosses and the chest of gold to prove his story. Congratulating the boy for his bravery, they told him to choose whatever money or goods he liked by way of reward and invited him to be their guest at Skalholt for as long as he wanted. He thanked them but declined their offer, saying he planned to leave in the morning. That night, everyone once again slept at the house, with no sign of trouble from the ghost.

In the morning the lad gathered up his things, in spite of the folk's earnest entreaties of him to stay. He was no longer needed, he told them. And so he left Skalholt and traveled north to the summer grazing pastures.

For some time after he arrived at this new locale, nothing unusual happened. But one day as he was exploring, he came upon a cave. Going inside, he found it was deserted. But in a side chamber, he found twelve beds—six on one side and six on the other. The beds had not yet been made, and since it was still daytime and the lad figured their owners probably would not be back for a while, he went about making them. Then he lay down in the bed that was the furthest on the right, covered himself well, and fell asleep. After a while, he was awaken by a great commotion from a group of men who entered the cave and began conjecturing as to who had done them the service of making their beds. Whoever it was had earned their thanks, they said. Thereupon, they ate their supper and prepared to retire for the night. But when the man whose bed was the furthest on the right threw back the covers, he found the lad lying there. Now all the men joined in thanking the lad for his services and pleaded with him to stay with them. They explained that they had to leave the cave every morning at sunrise or else their enemies would find them and attack them. That was why they never had time to do housework and would be ever so grateful if the lad would consent to stay on and help them with their chores.

The young man agreed to stay with the cave dwellers. But he was curious about one thing. How was it that they got into such bad fights day after day? The cave dwellers explained that the battle with their enemies had been going on for as long as they could remember. Though they had always been the victors, lately their foes kept coming back to life in the morning, re-newed and even more full of wildness and malice than before. Surely the cave dwellers would be attacked in their own home if they were not up and ready to fight at sunrise, they explained. Then they lay down and slept until the next morning.

When the sun rose, the cavemen armed themselves and went out, but not without asking the boy to look after the cave and do some housekeeping while they were gone. Though he cheerfully agreed, as soon as they had left, he followed after them into the forest of walnut trees where the fights took place. Once he had surveyed the battlefield, he quickly hurried back to the cave where he made the beds, swept the floor, and took care of all the other chores.

When the cave dwellers returned that evening weak and weary from their battles, they thanked the boy for taking care of everything so well. After eating their supper, they fell asleep immediately—all except the boy, who lay awake thinking about what might happen if the cave dwellers' enemies were once to attack at night. Thus, when he saw that his hosts were all asleep, he got up, selected a few choice weapons, and left the cave. He arrived at the battlefield just before midnight, but saw nothing except slain bodies and severed heads.

After a while, however, he saw a woman emerge from a hill not far from the battlefield. She was wearing a blue coat and carrying a can. The lad watched as she went straight to the battlefield, knelt beside one of the slain warriors, and wiped something from the can onto the upper part of his torso and the bottom of his neck. Then she placed the head back on the body. Immediately, the dead man sat up and came alive! Aha, thought the lad. Now he understood how the cave dwellers' enemies kept coming back to life. The woman went on to revive two or three other bodies in the same way before the boy lunged at her and struck her down. Next he slew the men that the woman had brought back to life and experimented to see if he could do the same. Indeed, wiping some of the slime on their necks, he found that the trick worked just as well for him as it had for her. For the rest of the night he continued playing this game, reviving the dead men and then killing them off again.

When they awoke at dawn, the boy's cave-dwelling friends were dismayed to find the boy missing along with their best weapons. But when they arrived at the battlefield, prepared for the day's fight, they were cheered to find their enemies still lying dead on the ground! When they saw the young man, they greeted him with joy, asking him what had brought him there. They boy told them everything that had happened. He showed them the can of ointment the elf-woman had used to revive the slain men and, wiping some onto one of the dead men, put his head back on. As quickly as the slain man rose up, the cavemen cut him down again.

Now the cave dwellers thanked the lad profusely for his bravery, bidding him to stay with them as long as he wanted and offering him money for his good deeds. He gratefully accepted their invitation.

This made the cave dwellers so happy that they began celebrating rowdily. Since they now had the power to bring one another back to life, they decided it would be fun to see what it was like to die. And so they merrily went about killing each other, applying the magic ointment, and coming back to life. They amused themselves in this manner for quite a while.

At one point, they cut off the boy's head and put it on backwards. When the lad revived and caught sight of his rear end, he went mad from the horror of it. Finally, he had encountered something truly frightening! Pleading to be released from this torture, he promised his liberators anything in the world. Straightaway, the cave dwellers came to hack off his head and put it back on the right way. When the boy came to his senses again, he was just as cheeky as ever.

Now the friends dragged all their dead enemies into a pile, robbed them of their weapons, and burned their bodies, along with that of the elf woman. Then, going into the hill that she had come from, they stole all the treasures they found there and carried them back to their cave. The young fellow remained with them thereafter, and there are no more stories to tell about him.

This fairy tale has a direct parallel in the Grimms' tale, "The Boy Who Left Home to Find Out about the Shivers,"[2] and variants of it have been found throughout the world. Though the hero's deeds are not identical in all versions, the great lengths to which the tales go in describing them are a constant. After reading a number of these tales, one gets the feeling that the teller quite enjoys identifying with the saucy hero. Thus we may assume that this is a character type that embodies some sort of ideal, even though the tales make it clear from the beginning that lack of fear is actually a defect and thus that the boy is inhuman, or better, prehuman. These tales describe a developmental process that, though it culminates in the protagonist learning to feel anxiety and horror, never requires him to lose his pluck. In many versions, the protagonist begins a relationship with the woman he wins through his bold deeds—despite the fact that this seems a detour from his primary goal, namely to learn fear. Thus, from the perspective of the tale, there is a close connection between the ability to feel anxiety and the ability to enter into relationships.

What sort of growth process does the lad whose goal is to experience anxiety and horror undergo? At the beginning, the tale gives no indication that the boy himself feels at all concerned about his lack of anxiety; it is only his parents and relatives who are worried about it. And what seems to upset them most is how the lad's fearlessness deprives them of authority over him.

In their desperation, the parents send the boy to the pastor. In former days, this was a popular option: When parents were no longer able to deal with their children, they would often send them to the pastor to be straightened out. On a superficial level, the tale does portray a need for authority. It is the pastor who is supposed to find a way to instill fear in the boy. Fear is thus connected not only with authority, but also with religion and the divine. The pastor's task brings to mind all that is meant by the expression "the fear of God." It is associated with the individual's need to find his place in relation to the *numinosum* and to human life as a whole. Luther, for example, began his interpretation of the Ten Commandments with the sentence, "We should fear and love God. . . ." A boy like our hero, who lacks even the slightest hint of fear or sense of danger, will identify with all things tremendous and omnipotent. Thus the pastor might help with this problem by demonstrating a healthier relationship with the transcendent than the boy's direct identification with it.

But in our tale, the pastor fails at this task. Thinking that a confrontation with mortality will shock the cheeky lad out of his fantasies of omnipotence, the pastor sets him up to stumble over a set of corpses. And indeed, in the end it is the awareness of transience and mortality that gives the boy his first taste of horror.

However, though the encounter in the church may be a first step toward knowing fear, at this point the desired shock does not take place. Indeed, the young man deals with the dead as if they were wooden puppets, showing them no sensitivity at all. Perhaps it is not mere coincidence that this part of the tale takes place in winter. The temporary hiddenness of life at this time of year suggests a repression of anxiety and other feelings, such as respect for the dead.

The episode with the pastor marks the boy's initial attempt to deal with the problem of death. Afterwards, he is sent away—which, in the language of fairy tales, means that he must continue in his growth process.

What sort of person hides behind this fellow's bold facade? The tale describes him as saucy and brash, but also as surprisingly helpful, playful, curious, and indomitable. If he is hungry, he finds something to eat; if he craves adventure, he drums it up. Fearing nothing himself, he is feared by those around him, who cringe at his lack of self-imposed limits.

One function of anxiety is to make us conscious of our limits. It is only sensible, for instance, to have some trepidation about climbing a steep mountain, since one could easily take a fatal fall.

The lad in our story regards himself as omnipotent. This is not to say that he wants to lord his power over others—at least not consciously—only that he believes the world is at his fingertips; he can trust in his own powers to obtain whatever he wants and needs. It was this uncritical, indeed naive, trust, expansiveness, and power that "restored the courage" of the folk of Skalholt.

I find myself intrigued by this indefatigable youth. With his sense of freedom and power, he inspires "courage" in me, too. But he also frightens me. Perhaps it is the tale's hidden agenda to affect its audience in this way. By portraying a character who is entirely free of anxiety, it awakens in us the desire for a life of fearlessness, boldness, brashness, for the thrill of being swept along by a steady stream of good luck. And yet at the same time, the character evokes in us the opposite of boldness—namely fear.

Let us regard the hero of this tale as someone who needs to take a significant step toward greater maturity. Unaware of his limits, convinced that he will never fail, and possessed of great strength, the lad is the picture of a slightly manic personality. Although he seems lucky, he is highly restless; his being lacks some essential dimension. Stated another way, he is someone who is protected—and trapped—in a positive maternal space. Thus enveloped and blinded, he represses consciousness of the dark side of the maternal realm: the womb as a place of death. People whose lives are contained in this way often suffer from boredom and a sense of emptiness, especially when no one is around to admire them for the allure they possess by virtue of identifying with the merely good mother. Such people lack a sense of the transience of life, and consequently a sense of the weight and importance of the present moment.

Thus the hero of this tale will be sent on a journey to confront religious figures and death. His next test takes place in Skalholt.

"Skalholt" is a name rich in symbolism; translated into English it means "Skull Hill." When the tale announces that the bishop has died, it is clear that we are about to encounter the mystery of death and the next world in yet another form. The bishop represents an "authority" even "higher" than the pastor. Ever since the bishop's death, we are told, the night has been haunted. In mythology, a figure "haunts" a place when it can find no peace, but if someone recognizes its need, the ghost usually can be redeemed. Such haunting figures can be seen as aspects of the unconscious that, having been repressed at some earlier time, have again become activated and now press for recognition by the conscious mind. For example, we say that "something is haunting me" when we feel disturbed, confused, and preoccupied by an energy that we cannot define more precisely.

The ghost in our story makes its appearance after the lad cuts up some dried meat and puts it in a pot to cook. The boy wants to eat, and there is plenty of food available in this well-stocked kitchen. Yet his "appointment with the ghost" prevents him from feeding himself.

In fairy tales, it is quite conventional for ghosts to enter a dwelling by means of the chimney. Witches, too, go in and out of houses through the chimney, as do the souls of the dead. This may have something to do with the resemblance between chimneys and caves, a frequent symbol of the realm of shades or souls. The blackness of the chimney suggests the same thing. Fire and, especially, rising smoke are also associated with the realm of the spirits. As for the hearth, it not only symbolizes the center of the family and human society, but transformation.

In our tale, the ghost that falls down the chimney does in fact effect a transformation. Its three pieces recall the three corpses in the church from the first part of the tale. However, by telescoping three separate bodies into one body in three pieces, the tale suggests that the story's central problem is now becoming more focused. And the fact that these pieces land squarely in front of the protagonist rather than allowing themselves to be shoved aside suggests that the problem has also become more immediate and compelling.

One might imagine that the young lad's confrontation with the corpses in the church was so disturbing that it caused him to completely suppress any sign of anxiety. In this regard, the encounter could be compared with particularly penetrating interpretations made in therapy. For example, if, in a submanic mood, the therapist prematurely identifies a fear of mortality in the patient, the emotional impact may be such that the patient completely represses it. But just as the corpses were shoved aside only to reappear later, so is the insight that caused such anxiety in therapy gradually admitted to consciousness.

It is striking to see how the boy deals with the dismembered body—as if it were nothing out of the ordinary. He does not experience the ghost as dangerous, intrusive, or alien; he even encourages it to pull itself together so that it can regain its mobility. Here we see the advantages of a naive attitude: Exempt from the need to defend himself, the lad can look on with interest, watching what happens. But we can also see the drawbacks of such an attitude: Lacking any sense of danger, the boy may well lose his life in the battle with the ghost.

But, before we go any further, let us pause to inquire into the nature of this ghost. What sort of problem does it represent? What psychological complex waiting to be resolved and redeemed?

Let us start by saying that the ghost's actions reveal something about the nature of the problem while the boy's actions—his dealings with the ghost—reveal something about how the problem can be tackled. The ghost is a terribly large man who behaves autonomously, even obsessively, following a kind of "blind will." In his article "On Psychic Energy,"[3] Jung described spirits as autonomous complexes that have not yet been integrated into consciousness. These complexes reveal themselves if we make them the subject of active imagination—for example, when we draw or paint a picture of whatever is haunting us, our suspicions, tensions, or longings. Likewise, in our story the problem of anxiety and death is portrayed in a series of symbolic gestures enacted by the huge man. These involve entering a room where there is a chest full of money, removing all the coins and throwing them over his shoulder, then turning around and reversing what he has just done, throwing the coins back into the chest.

The "ghost that counts money" is a frequent motif in folk literature. It illustrates the insight that those who grasp at money for security in their lifetimes go on grasping at it when their lives are over. Such a motif might be interpreted as advice to take leave of your money now if you do not want to be counting it throughout eternity. Our tale vividly illustrates the curse of such a fate. The ghost's act of throwing money over his shoulder and then gathering it up again is indeed a Sisyphian one.

If we examine this motif of throwing something away in light of its occurrence in other tales, we gain the impression that it has to do with sacrifice. For example, in a number of tales, the hero on a magical flight throws behind him or her a series of objects which then grow to huge proportions—for example, a bush turns into a forest. The sacrificed object safeguards the person in flight and may even save his or her life.

In our tale, it is not so clear that the ghost's throwing away money connotes a sacrifice. Indeed, the failure to make a genuine sacrifice seems to be a major cause of the curse that pervades the entire tale: though death has forced the bishop to give up his money—and power—by dying, apparently he was never prepared to surrender these willingly. Thus, nothing has really been sacrificed—as can be seen by the ghost's obsessive gathering up money after he has dispersed it. This image shows what the bishop (I believe that the ghost and the bishop are one) would have had to do to fulfill his religious duty, accept his mortality, and become fully human: sacrifice his money and power. The curse results from his inability to do so.

The "will to have" expresses a fear of loss, change, and finally, incapacitation. As such, it can be seen as a defense against "transience" (Weischedel[4]). Everyone must die, and even while we are living, death continually encroaches on us via the transience of life. To face life's transience without becoming resigned, we need creativity, which is the will to shape our own being. In my view, the "will to have" and its associated "will to power" make up the shadow side of creativity.

Applying these insights to the psychology of the individual, we might say that those who strongly repress their anxiety—or who simply never grew out of their infantile feelings of omnipotence—lack an appropriate sense of danger. Their style is submanic if not hysterical: bold, full of good cheer, naive. Even so, at some point they notice that something is missing—sorely missing. They may realize this themselves or it may be pointed out by others, but suddenly they will encounter major problems with power and the will to possess, and also with the necessity to yield to others. These problems will present themselves with the force of an assault. Notice that it is not until the lad in our tale realizes he is hungry—i.e., that something is missing—that the ghost makes its entrance.

How does the lad deal with this problem—with this "complex"? To begin with, he admits the ghost into his sphere of consciousness. By first paying attention to its separate parts, he allows the spirit to "assemble" itself so that it can make its secret known. The young man then follows the ghost about, observing its obsession, allowing it to give its own account of itself. After learning the ghost's secret, the lad engages it in combat.

Through the ensuing confrontation, true growth in consciousness takes place.

To me, this passage in the tale reads like an instruction manual on how to deal with an overpowering complex: First, allow it to give an account of itself, watching where, when, and how it functions; then attempt to engage it in an encounter, taking care not to let it retreat into the dark of night, for that would mean forfeiting it to the unconscious again.

The encounter requires great cunning from our lad. This means he must learn his opponent's tricks and anticipate his moves; he must identify with his opponent but also regard him from a distance. In the situation described by the tale, the problem is undone once it has been unmasked; after that, it no longer pins one to the wall, one is free to assert oneself. Although the boy realizes that the fight with the ghost could cost him his life, he is not paralyzed; he acts—and yet I do not have the impression that he still fancies himself—immortal.

The outcome of the fight is that the ghost splits into two pieces. This happens at daybreak—which is to say that once light has been thrown on the problem it ceases to exist, at least in the old form. In the story, one half of the "problem" sinks back into the earth in broad daylight while the other half sinks into the floor inside the house, suggesting that some inward, psychological aspect of the problem has yet to be resolved. This aspect is associated with the lower half of the giant man who, as we may recall, came down the chimney "head first." The "underside" of the problem—corresponding to the man's organs of digestion, excretion, and sex—has yet to find peace. The "lower half" of the body is a most powerful symbol of our mortality, reminding us daily that we exist within a cycle of holding and letting go, that all things pass and we ourselves are transient.

Why should the bishop be the symbolic figure in which our hero confronts the problem of holding on and letting go? Why do I maintain that the ghost and the bishop are, for all purposes, one and the same?

First let us consider the collective, historical conditions that might have given rise to this tale. It is conceivable that the story was created in a time marked by a general flouting of death and mortality, a penchant for high living, and an open preference for power and wealth over other worldly concerns. Perhaps the tale expresses disdain for those whose calling it was to cultivate a relationship with the next world but who succumbed to the lures of this one. It would be hard in my view to attribute such conditions to a specific historical epoch, since the church has been plagued by such abuses throughout its history.

If, instead of looking to historical factors to explain the tale, we view it as expressing the inner parts of an individual personality, we can see the bishop as an authority figure. Then the ghost would represent a repressed authority complex, such as we often encounter in people who claim to recognize no authorities. Concealed behind this authority and power complex we are likely to discover a religious problem.

Having put half of this authority problem to rest—but leaving the "underside" still to be resolved—the boy now moves on to the summer grazing pastures. Here in life's "midsummer" a "night's dream" takes place that effects a decisive change in the lad's existence.

The lad finds a cave. Caves offer retreat into a protective space, an inner, maternal realm. This is why in various symbologies they are associated with birth. For the same reason, they have also been imagined as entrances to the realm of the dead. For example, the Sumerians told of a realm of shades in a cave in the World Mountain. Caves have also been very important in rites of initiation, such as the Eleusinian mysteries. This is probably because entering caves provides an impression of regressing, of reentering the womb and reemerging in a second birth. In caves, we can imagine dying and being reborn, and thus they are also the symbolic sites of fundamental changes in the personality.

Elemental transformation is likewise expressed in the image of the twelve beds. The number twelve suggests wholeness (it is the base number of the Babylonian duodecimal system). As such, it is a symbol of completion, signifying the end of a spacio-temporal cycle. In twelve hours or twelve months, a cycle completes itself. But in the tale, the number twelve is divided in two—six beds on one side face six on the other. To me, this suggests that although the entire life-situation is about to change and a new beginning is on the horizon, what lies in store may not be the birth of an integrated whole so much as a confrontation between two poles of reality: six facing off against six.

The boy shows that he accepts this situation and, by making their beds, prepares the way for the cave dwellers to find peace. His helpfulness is then praised with great enthusiasm.

The cave dwellers are in great need of someone to solve a long-standing problem. Every morning, they go out to fight their enemies. In the evening, they declare themselves the winners, only to find in the next morning that their vanquished enemies are alive again and wilder than ever. Just like the ghost of the bishop who repeatedly demonstrated his inability to let go and the huge ghoul who was not really finished off when it split into two pieces, here is something that refuses to die and declare itself defeated. Here is another "eternal" conflict.

If something cannot die, then it cannot be "buried"—it cannot be forgiven and forgotten. In such a situation, a conflict between hostile factions is perpetuated ad infinitum. I think here of street gangs and their never-ending battles; even when one gang has "won," the members find no peace and quickly resume their feuding. They usually do this quite unconsciously, perhaps responding to a lurking feeling that the winner is not a winner outright, but also to some extent a loser.

What is at the root of this eternal conflict in our tale? What causes the slain men to keep coming back to life? What thing or person is at work behind the scenes?

It is, we are told, a woman in a blue coat who emerges from a hill carrying a can of ointment for reviving the dead. Though the tale identifies her as an elf-woman, we know that generally in Icelandic tales, women who emerge from hills are fairies. What do we know about fairies that might explain her role in this conflict?

According to the *Dictionary of German Superstitions*, fairies come from a "slightly exhilarated Celtic-French fantasy world." Embodiments of celestial virtue, they sometimes come to the aid of persons undergoing severe trials. But just as often, they simply dance in their grottoes with the joy and abandon of *eternal youth* "fully adverse to the concept of death."[5] As agents of the principle of good, they rarely perpetrate evil deeds—and these only when they are meting out justified retaliatory blows to ungrateful persons. The fairy must stay "light" in every sense of the word. For example, the fairy in our tale wears a blue coat, which imbues her with the aura of the Virgin Mary and emphasizes her celestial nature. Fairies come from a land of many marvels, bringing such miraculous objects as magic hats and rings that, when turned, transport the wearer from one place to another. These marvels come to the aid of severely mistreated persons, reminding us of the miraculous therapeutic potential of the creative imagination. How often has it happened that, when caught in a difficult situation, we immerse ourselves in a fantasy, and emerging, find that what was hitherto unimaginable has suddenly become possible?

But how does this explain the role the elf-woman plays in perpetuating the bloody conflict in our story? This fairy is, in fact, a dangerous one—at least for the hero of our tale. At the beginning of this essay, I said that the tale describes the psychology of a "merely good" maternal sphere. If we include the phenomenology of the fairy world in this description, we can begin to understand some of the danger such "goodness" presents. This fairy world is too light. It is intoxicating—not in an earthy, Dionysian way, but rather in the way of a tantalizing illusion. Now we can see that behind the will to possess and the inability to yield there lies the fantasy of everlasting life and eternal youth. We saw it in the figure of the bishop and in the image of the enemies who kept coming back to life. In each case, something was not allowed to be lost and forgotten.

As they are described in the literature, fairies are simpleminded, insufficiently embodied, and unreal. This is what makes them so refreshing for those who are torn between opposites and bound by the limits of reality. For those who do not feel the conflict of opposites or the limits of earth, the fairy may mean something else altogether. Indeed, for the boy in the tale, the fairy represents a dangerous possibility; the ideal that she embodies perpetuates an eternal clash within him.

Perhaps the boy understands this danger, for he kills the fairy. Could it be that he has finally outgrown an attitude that sees everything as a game and does not recognize death? This seems unlikely, because after executing the fairy, he plays one last, spirited game of killing and reviving, showing no sign that he has glimpsed anything of the horror or finality of death. Together with the inhabitants of the cave, he indulges in an orgy of death and resurrection, a manic game lacking all sobriety. Although the tale is told in a style completely devoid of feeling, it has a paradoxical effect on the reader—horrifying us precisely with its lack of horror.

In a way, this tale is reminiscent of certain adventure films or Westerns in which the characters die like flies. Since these are "only" films, we viewers hardly take the killing seriously, knowing that the actor survives in spite of appearances. I suspect that such films appeal to a haughty, omnipotent attitude within the viewer that fancies itself sovereign over life and death. No one is completely immune to this attitude, which is quite prevalent today in our collective culture. I think not only of the "dying and rising" games of fairy tales but also the suicidal fantasies of certain persons who play with the idea—and reality—of killing themselves. To these people, life and death become a game whose shock value registers with others but who remain themselves oblivious to death and its irreversibility.

By killing the fairy in the blue coat, our lad seems to demonstrate a need to stop denying death. However, this does not mean he can easily divorce himself from his gaming mentality—he goes on killing wildly in accord with the general principle that behavior patterns often get much worse before they get better.

The hoped-for change does take place, however, when the lad's head is cut off and then put on backwards. When the boy sees his own rear end, the story tells us, he suddenly goes "mad from the horror of it." Thus we might say that the cave dwellers "turned the boy's head around"—altered his perspective—and thereby gave him the shock of his life. Although his own behind may seem a banal sight compared to all the killing that has gone on before, there are good reasons for it to have such a horrifying effect on the lad. Indeed, the tale suggests that only when we recognize the implications of this lowly part of our being can we come to terms with our mortality. Seeing our own backside reminds us that we are but creatures, animals. All that is corporeal suffers the fate of being taken in only to be expelled again. Everything is subject to decay and death. Ultimately there is no escaping transience, finitude, and danger.

Mad with horror, the boy asks for help for the first time. No longer the gleaming hero who comes to everyone else's rescue, now he is the one in need. Neither is he oblivious to the limits of the body and of reality any longer. Conscious—at last—of his own mortality, he has finally become a human being.

Let us imagine once more an individual who knows that death is a part of life and yet plays manic games with this knowledge. In principle he knows that all is transient—at least as far as others are concerned. And then one day while sitting on the toilet, he is struck by the realization that we can never hold on to what we have brought into our lives. We are but clay vessels; even our most sublime visions are born out of these earthen jars, which are subject to the wear and tear of sun, rain, and the passage of time. This realization plunges him into an abyss of depression, wherein he watches certain of his illusions die while at the same time discovering new depths that reach to the painful bedrock of human reality.

Our tale indicates that, following his abysmal realization, the boy continued on as boldly as ever—taking all the riches out of the fairy hill to deposit them in the home of the cave dwellers. This seems to hint that even the most one-sided attitude contains a treasure with potential value for the future. Once he invests himself in that future, the boy can at last put his dead enemies to rest.

That "there are no more stories to tell about" the boy is to be understood as a narrative device for closing the tale. The tale promised to show us the development of someone who was never afraid—not a normal, properly anxious human being. Thus it ends when the lad discovers that he is mortal and that anxiety and threat are an ineradicable part of human life. This "knowledge" seems to be a prerequisite for real relationship. At least that is the suggestion of those variants on this fairy tale in which the defiant hero goes to live with his wife once he has finally discovered what horror is.

■ The Goose Maid

Once there was a queen whose husband had died many years before. The queen had a beautiful daughter, of whom it was promised that, when she grew up, she would marry a certain prince who lived far away. When it came time for the girl to be wed, her mother packed many precious things for her journey to the foreign land—jewelry, gold and silver, goblets and treasures, in short, everything belonging to a royal dowry, for her mother was very fond of her. Last of all, the queen provided her daughter with a chambermaid who would ride with the bride and deliver her into the bridegroom's hands. Both girls were given horses for the journey, but the princess's horse, whose name was Falada, was capable of speech. When the hour of departure came, the old queen shut herself in her bedroom, took a small blade, and cut her finger to draw some blood. Then she let three drops of the blood drip into a small white cloth, which she gave to her daughter. "Dear child," she said, "take good care of this; you will need it in your travels."

Sadly, mother and daughter said their goodbyes. The princess tucked the cloth in her bodice, mounted the horse, and rode off to meet her lover. After an hour's ride she became very thirsty and said to her chambermaid, "I'd like a drink. Get down and scoop me a drink of water with the cup you brought for me." But the maid replied, "If you're so thirsty you can get down and get it yourself; I'm not your slave." So the princess, who was very thirsty, lay down by the brook and drank directly, because the chambermaid wouldn't let her use the golden goblet. "Woe is me!" said the princess, and the three drops of blood in the cloth answered back, "If only your good mother knew, her heart would surely break in two." But the princess was meek; she got back on her horse without saying anything.

Now the two girls rode along a good many more miles, but it was terribly hot and the sun burned their skin, and soon the princess grew thirsty again. Arriving at another brook and, having completely forgotten the lady's response the first time, she called once more to her chambermaid, "Get down and give me a drink from my golden goblet." This time the maid's response was even shorter. "If you want to drink, go ahead and drink; I'm not your slave." The princess was so thirsty that she did not even hesitate, but got down off her horse directly, lay down by the rushing water, and cried, "Woe is me!" Once again, the drops of blood in the cloth answered, "If only your good mother knew, her heart would surely break in two!" As she leaned out over the water to take a drink, the cloth fell out of her bodice and was swept away with the stream. But in her anxiety, the princess did not even notice. The lady in waiting noticed, however, and gloated secretly. Without those drops of blood, the princess was weak and powerless. So when she went to mount her horse, the maid said, "Falada's mine now; you can have my old nag," and the princess had no choice but to accept her harsh pronouncement. Then the nasty maid ordered the bride to take off her royal garments and exchange them for the old rags she herself wore, and to swear to the heavens above that she would not say a word of this to anyone at the royal court. If she refused to take this oath, the maid said, she would be killed on the spot. But Falada saw everything and marked it well.

Once the chambermaid had mounted Falada and the true bride had gotten onto the old nag, the two traveled on to the royal palace. When they arrived, they were greeted with great jubilation. The prince ran out to meet them and helped the chambermaid down from the horse and, thinking he held his spouse's hand, led her up the stairs. The true princess had to stay in the courtyard below, where the old king, from his window, saw her looking very gentle and fair. Right away he returned to his royal chamber and asked the false bride about the girl who had accompanied her. "Oh, her. I found her along the way and took her along to keep me company. Give her some work to do so she'll stay out of trouble." But the old king couldn't think of anything, having no work in need of doing. "I have a young boy who tends geese," he finally told the true bride. "Maybe you can help him." The boy's name was Conrad, and the true bride went to help him tend geese.

Before long, the false bride asked the young king, "My dearest husband, I beg a favor of you."

"But of course. What is it?" the prince replied.

"Send for the slaughterer and have him hack off the head of that horse that I rode here on. It was a bad horse and made me angry." Actually, she was afraid the horse might say something about what she had done to the princess.

The lady's request was granted, and arrangements were made for faithful Falada to be killed. But the true bride got word of it and promised the slaughterer a sum of money if he would do her a service. In the city was a large, dark gate through which she passed every morning and evening on her way to tend the geese. The princess asked him to hang Falada's head from a nail over the gate so that she could see him twice each day. The slaughterer gave her his promise, hacked off the head, and nailed it at the top of the dark gate.

Early the next morning, as she and Conrad drove the geese through the gate, she addressed the horse's head, "Oh Falada, hanging there!"

And the head answered back,

"Oh, young Queen, how ill you fare!
If only your good mother knew,

Her heart would surely break in two."

The princess made no reply. She simply went out through the city gate and drove the geese out into the pastures. When she reached a meadow, she let down her hair, which was of purest gold. Its glistening radiance awakened Conrad's desire and made him want to pull out a few strands for himself. Perceiving his designs on her, the princess spoke the words:

"Blow, ye gentle winds, I say,
Blow Conrad's little hat away,
Make him chase it here and there,
'Til I have braided all my hair
And bound it up again."

And there came such a strong wind that it blew Conrad's hat into a distant pasture and sent him chasing after it. By the time he was back, the princess had finished combing and putting up her hair and Conrad wasn't able to pluck a single strand. This made him so angry that he refused to speak with her. In silence, they looked after the geese until evening, when they returned home again.

The next morning when they drove the geese out through the dark gate, the girl again addressed the dead horse:

"Oh Falada, hanging there!"

And Falada answered back:

"Oh, princess, how ill you fare!
If only your good mother knew,
Her heart would surely break in two."

Once they arrived in the meadow, the princess started combing out her hair again. This time, Conrad leapt up immediately, attempting to pluck a strand or two, but the princess quickly called out,

"Blow, ye gentle winds, I say,
Blow Conrad's little hat away,
Make him chase it here and there,
'Til I have braided all my hair

And bound it up again."

Immediately the winds came up and blew Conrad's hat far away, and he went running after it. By the time he returned, the princess's hair was neatly arranged on her head with nary a strand left for him to pluck. The two of them tended the geese until evening.

But that night when they returned home, Conrad went to talk to the old King. "I refuse to tend geese with that girl any longer," he declared.

"But why?" asked the old king.

"Ah—she irritates me the whole day long," said the lad. Then the old king bid him to describe in detail all that transpired during the course of the day. "In the morning," Conrad began, "when we take the herd through the dark gate, she talks to a horse's head hanging from the ceiling. 'Oh Falada,' she says, "hanging there.' And the head replies, 'Oh princess, how ill you fare. If only your good mother knew, her heart would surely break in two.' Conrad went on to relate everything that happened in the pasture where the geese grazed, including the winds that daily blew his hat away and sent him running after it. After listening to all that he had to say, the king ordered Conrad do everything the next day exactly as before. As for the king, he hid behind the dark gate and heard for himself the goose maid's strange greeting to the dead horse's head. Then, following her out into the pasture, he hid behind some bushes. With his own eyes, he watched the goose girl sit down and undo her hair, and saw how it glistened. He heard her say,

"Blow, ye gentle winds, I say,
Blow Conrad's little hat away,
Make him chase it here and there,
'Til I have braided all my hair
And bound it up again."

He witnessed the gust of wind that took Conrad's hat with it and watched the boy run off into the distance while the maid sat calmly combing and braiding her hair. Still unnoticed, the king returned to his palace. But that evening, when the goose maid returned, he summoned her to his chambers to have a talk with her. Why did she do what she did, he asked. "That I cannot tell you, nor anyone else, for I swore it to the heavens above, and breaking my oath would cost me my life." The king tried everything he could think of to entice more information out of her but she refused to say anything more. And so the old man, who was wise, said, "Well, if you won't talk with me, maybe you can let the old cast iron stove know what is troubling you." And he turned and walked away. Indeed, the true princess crawled right inside the stove and began weeping, moaning, and pouring her heart out. "Here I sit, abandoned by the whole world—and I'm the true princess. My own chambermaid tricked me, took me by force, stripped me of my royal garments, and stole my husband before my own eyes. Now I am reduced to the miserable job of tending geese. If only my good mother knew, her heart would surely break in two." All the while, the old king was standing outside, listening to her confessions through the stove-pipe. Now he came back in and commanded her to climb out of the oven. Without further ado, he had her fitted with royal robes, letting her beauty shine forth miraculously. Next, he summoned his son and informed him that the girl he was planning to marry was the chambermaid. His real bride was standing before him, previously a goose maid. When he saw her beauty and virtue, the young prince was overcome with joy. Soon, a magnificent feast was prepared, to which all courtiers and good friends were invited. At the head of the table sat the bridegroom, with the princess on one side and the chambermaid on the other. The maid was confused, however, and did not recognize the bride anymore because of the sparkling jewelry and royal robes she wore. Once they had had plenty to eat and drink, and everyone was in jovial spirits, the king described for the maid a hypothetical situation. What

ter—for example, the family. In this case, however, it would have to be a family that lacked something, perhaps one where the unspoken rule was that all aggression would be transformed into intensified caring for one another's needs. A current trend demands that daughters should emancipate themselves from their mothers. Rather than accepting inherited values and images of women unquestioningly, daughters today are expected to be more aware and independent.

In our story, the mother-daughter relationship appears to be a close one, and the princess is endowed with much inherited wealth. Still, it is the queen herself who initiates the separation, by promising her daughter's hand to the prince. Even if she is "very fond" of her daughter, she does not want to keep her. What does the mother give her daughter to take on her journey? A royal dowry, including gold and silver. The princess thus is equipped with things of great exterior value. As yet, however, the daughter has not earned this worth for herself; it is only an inheritance.

The mother also gives her daughter a chambermaid. While, throughout the tale, the princess is presented as the "good" one, and commands our sympathy, the maid is always portrayed as the "bad" one, stimulating our enmity and dislike.

It is unlikely that the old queen deliberately sent an evil chambermaid with her daughter. More likely, badness in general has been omitted from the mother-daughter relationship. Growing up in a sheltered maternal environment, the princess has had few opportunities to deal with the "chambermaids" of the outer world—that is, persons who would try to gain power over her, who would like to steal her privileges and her place. Nor has she had to deal with the "chambermaids" of the inner world—including such inner psychic phenomena as power-drives, deceptive intentions, callousness, etc.

But along with the chambermaid, the mother has sent with her daughter a horse that is able to speak: Falada. According to the research of Schliephacke,[7] Falada means "sanctified in the name of the old god of light." Thus, the horse is associated with Wotan, and the dark gate over which his head hangs can be seen as an exit to the nether world. One may recall that Wotan hung for nine days on the windy tree and spoke to a head in order to know himself better. The Germanic peoples had a tradition of fastening the skulls of horses in the gables of their houses in order to gain Wotan's help. Wotan was a god of horses, wind, and spirit. In our folktale the wind plays an important role as well; the princess knows how to deal with the wind, indeed, the wind is at her command. Horses symbolize, among other things, a person's instinctive, animal nature, the dynamic energy of the body.

The connection between horse and wind is suggested by the way both can go wild. Ordinarily, the symbol of the horse expresses vitality. But in the image of a horse that can speak, we also find "wisdom" portrayed. This is the case in various other folktales in which horses speak. For example, Tacitus wrote about the Teutonic practice of consulting horses for prophecies, due to a belief that horses have a special rapport with the gods.[8]

If we see the princess as a daughter who needs to separate from her mother, I think we can rest assured that at least with Falada she has a good instinctive foundation and relationship to her body. She is energetic, senses changes within her body, and can deal with the unconscious. But the horse that is sacred to the old god of light also represents a problem. If the horse is seen as an allusion to Wotan's exclusively inspiring aspect, there may be too great a demand for light. As an image of the masculine, the horse may represent confinement to an overly ideal realm.

But the princess has also received a white rag with three drops of blood on it. To begin with, this rag symbolizes the ongoing relationship with her mother. Even after the daughter has been separated from her mother, the relationship continues through the rag. The rag, which acts protectively when she is confronted with perils, could remind the girl of how it was when she was still with her mother. From time immemorial, blood has symbolized the seat of the soul and life-energy. The rag with the blood thus symbolizes a psychic lifeline to the mother, which has many dimensions. It provides comfort and safety, but it also constricts the princess's freedom of movement. At the same time, the talisman may also portray the goal of the daughter's journey: to synthesize "white" and "red"— i.e., marry purity and virginity with blood, passion, suffering, and sexuality. Of course, through marriage, the girl is connected with her mother in a different way, because certainly the latter could not have become a mother if she had remained pure and white. (White may also at times represent new beginnings and the as-yet indescribable).

But above all, the rag appears to me as a kind of maternal comfort blanket. Thus, the daughter is never completely alone so long as the drops of blood can reassure her, "If only your good mother knew, her heart would surely break in two." Strongly evocative of the mother bond, this oft-repeated saying reveals how little autonomy the girl has developed as yet.

As a result, the princess does not become enraged at the chambermaid or fight her fate; instead, she thinks lovingly of her mother, whose heart would break if she knew of her trials. For now, the princess need not be aware of the psychological import of her dependence on her mother; the narrative only requires that she feel comforted in her abandonment. Meanwhile, we who read or hear the story become conscious of the mother issue when we see how weak and subservient the daughter becomes once she loses the rag. Immediately, the chambermaid assumes a position of power over the princess, for the latter's strength was only to be found in the rag—in the assurance of her mother's help—rather than in her own autonomy.

For a girl to become autonomous, she has to separate from her mother and find her own relationship to the inner and outer masculine. Our princess must have been well-mothered—perhaps a bit too well-mothered. Basically, she has been equipped with many values that serve her for now simply because she has inherited them. They provide her with plenty of energy and life and put her into contact with her instinctual nature. However, she still lacks a relationship to a human masculine figure. In its place she has a somewhat idealized way of looking at the world. Everything that does not fit in with an idealized worldview is split off, though it will of course become active when the separation from mother takes place.

Separation is usually a matter of stopping or quitting, or in the case of breaking with the mother, of allowing dimensions of oneself to emerge that had not been permitted to live in relationship with her. Here the loss of the talisman would have the function of reminding the princess of what she is besides the queen's daughter. Departing from the expectations of what she had been allows very different personality characteristics to come forth. The newly emerging characteristics may not be very nice, and may, in their departure from old expectations, create guilt as well. The result may be the emergence of an alter ego such as our folktale describes in the figure of the chambermaid.

The chambermaid gradually takes over in our tale. She begins acquiring power when she refuses the princess's request to fetch a drink of water. Here, the princess's thirst may well be a thirst for life—after all, it is what delivers her into the power of the chambermaid.

Earlier, I described the chambermaid as representing a cluster of personality traits consisting of a will to power, a craving for prestige, and a lack of consideration. In other words, the chambermaid may be viewed as the princess's "shadow," as described by C.G. Jung. In Jung's view, the "shadow" includes all of those psychic and behavioral possibilities that we do not live out because they are not compatible with our conscious attitude. The shadow can therefore be positive as well as negative. It is clear that any change of consciousness comes about because one has lived out some part of one's shadow. This is especially true in cases of separation such as our story presents. At this first stage of her journey, the princess seems rather pale and lost. She is humble, accepting whatever comes her way, but has become subservient to the chambermaid, a puppet of evil powers. In other words, life is in the shadow. Or, as Jung says, the shadow is what entangles us in life.

At this stage, the princess presents a good likeness of girls and women who have a relationship with their mothers that shelters and protects them. In some ways, such a relationship is a great gift, but it brackets out the masculine. Such girls or women who finally get up the courage to leave their shelters (which can also be provided by a relationship to a "maternal" man), are completely surprised when the world does not greet them with the friendliness to which they are accustomed, and deeply injured when they encounter brutality and maliciousness. Blind to their own craving for power, they easily fall into situations where they have to "bend down," as the tale puts it, to get themselves what they desire. Used to being spoiled, they are unprepared to get for themselves what they need to survive. They are used to being treated as important, which makes them very unpleasant and bossy in situations in which they lose their importance. Thus, we can understand the chambermaid as an aspect of the princess. The instant the latter leaves her sheltered life, she begins to experience the two sides of herself: the one who falters when she has to get herself a drink, isn't very worldly, and cowers before maliciousness; and the one who lives out the compensatory attitude of assertiveness and does not shy away from dark deeds. The maid is aggressive, dominating, intent on becoming important and powerful. Though she seems to be autonomous, this is really an illusion; her independence is contrived.

Viewing the folktale as the depiction of a woman's developing autonomy, we can observe that when aggression, will to power, and contrived independence take the place of the ego, the personality undergoes enormous changes. Persons who we think we know well can suddenly take on completely new characteristics. If these changes are not to our liking, we say that the person has degenerated. At times this assessment may be close to the truth. So great is the desire to break away from their earlier life that they begin to live out their "worst" side. This is often the case with very nice women who, suddenly feeling exploited and naive, make an immediate about-face to show their hitherto unseen hard and dominating sides. With the "chambermaid" dominating all, hardly anything is left of a woman's princesslike niceness.

In her haste to slake her thirst, the princess loses the rag with the drops of blood on it. "In her anxiety, she did not even notice," we are told. Typically, in our anxiety, we too forget that we are protected, which may have the consequence that our "protection" falls away, as does our sense of our selves and our strength.

What effect does this intense anxiety have in the folktale? If we put ourselves into the tale, we can imagine ourselves as a sheltered girl. For the first time in our lives, we are traveling to a distant and foreign land. A single person has been hired to accompany us, but rather than helping, she oppresses and humiliates us. Thus, not only are we alone, we are threatened and don't know what our persecutor might think of next. Even if we understand the chambermaid to be an aspect of ourselves, the sense of threat and aloneness does not disappear, because such an inner shadow quality is indeed frightening to the self. When anxiety seizes us, we are paralyzed in the province of life where we normally feel at ease. If we understand the chambermaid as a personality trait of the princess, such anxiety would be the product of a guilty conscience. The rag with the three drops of blood may be a source of anxiety if it some-how communicates the mother's disapproval and accusation: "If your mother only knew what you are doing, it would break her heart." Such anxiety can be helpful if it stops one from doing something that is foreign to one's own nature, but usu-ally it is destructive, preventing one from taking the risk to do something new and different. It keeps one in an infantile state of being bound to the mother. Thus, because of the anxi-ety which causes her to reach back for her mother, the prin-cess loses the rag. Not noticing that she has lost it, she falls completely under the chambermaid's power. This transition is apparent when she surrenders Falada, exchanges clothing with the maid, and promises not to say anything at the royal court. But, we are told, Falada "saw everything and marked it well."

Once the mother's protection—one's safety net and lifeline —is gone, a radical change of personality can occur, a new motivation can replace the guilty conscience that has hitherto driven the princess on her journey. The folktale's way of expressing this is to say that she may no longer ride on Falada. Previously I said that such a situation seems to indicate a "loss of instinct." On the other hand, if Falada expresses the principle of light, then the princess's deposition implies that she has to sit for a time on a "dark horse" or undergo a "dark phase." The harshness of her present situation is in proportion to the ease of her previous life at home. Yet another image for the change that comes over her personality is the exchange of clothes. Clothing expresses what we want to show to the world, facets of our-selves that we wish to display or have others see. When a person starts wearing a different kind of clothing, we realize that their relationship to the world has changed in a way that is important for others to notice. In the folktale, the change is not deliberate, but one over which the princess has no choice. It is not the result of free will, but rather of manipulation and destiny. With nothing left to pit against these, the daughter is overwhelmed by the dominating, authoritarian side of her personality.

To be overpowered in this way by shadowy traits is a very frightening experience, and one that results only when such traits have been rigidly bracketed out. In fact, the chambermaid belonged to the mother. So the shadow problem has a history to it, and the anxiety bound up in it may have multiplied with the generations. When the daughter swears on pain of death to remain silent about her subjugation to the maid, we see how intense her fear of the shadow problem really is.

What could this oath to heaven mean? First, it means that the princess swears to a higher power that she will not divulge the deception that has taken place and that has produced this split between the true and the false bride. She will not let others know of it. Her integrity in thus keeping a secret seems to belong to her true personality, expressing a steadfast part of herself that is bound to heaven rather than simply to the maid, even if it was the latter who required the oath in the first place.

The oath insures that the princess will not tell what happened, and it marks the awareness that she is beholden to a greater power. I see the maid's threat of death as a suggestion that without a perspective that includes the transcendent dimension, the princess would not be able to survive; she would become possessed by the inner chambermaid, with nothing remaining of her true personality. The oath protects her from this fatal development. She also seems to want to protect her horse (or her instincts), as we are told that Falada (or something within her) registers what is taking place. By taking the oath, the princess makes a conscious decision to tolerate the situation as it is currently, and thereby leaves open the possibility of future development. By the same token, it is important for someone who has fallen completely under the rule of a formerly repressed aspect not only to recognize it, but to distance herself from it to some extent, remembering that other possibilities exist as well. If we speak exclusively to the false self of a person in this situation, the person will not be able to admit these other possibilities, even though she may be inwardly aware of them.

The princess and chambermaid arrive at the royal court with Falada, where there seem to be exclusively men: the prince, the old king, and Conrad. Looking only at her clothing and noticing nothing amiss, the prince is overjoyed to receive the false bride, but the old king sees the shape of the person behind the clothes. Thus, though the initial encounter with the personal masculine has taken place, the woman is not able to show her true personality. Nor does the man notice it, for his relationship to the feminine is not differentiated either.

Viewing the folktale from the perspective of a girl or young woman, we can see that once she has separated from the maternal, a girl can form a bond with the masculine—whether through relationship with a real man or with the masculine aspects of her own psyche. Here, however, the one entering into the relationship is not the whole person, but only the dominating shadow side. Thus we are shown a picture of a relationship in which everything looks good on the outside, and yet in which essential aspects of the personality cannot be lived. There is need for inner development.

And so the true bride becomes a goose maid. It is the old king, who has some sense of who the princess really is and what she needs to experience, who sends her to tend the geese. In ancient Greece the goose, sacred to Aphrodite, was a symbol of love and fertility. At the same time, as a creature that churns about in the muck, the goose is an attribute of the Russian witch, who often has the feet of a goose or lives in a house that stands on goose legs. Thus the goose is associated with a "witchy" aspect of a woman and the "muck and mire" of physical embodiment. At this point in the tale, the true bride is sent to attend to eros, to the dark feminine, to love and sexuality. To "tend" something involves concentrating on it, keeping it from dissipating. Thus, the true bride must concentrate on love in its entirety, because whatever connects her and the prince at the moment clearly falls short of this. Perhaps this is a so-called "exclusive relationship" in which she feels obligated to remain faithful to the prince, but it is far from a real relationship. With the work of tending the geese, however, the princess begins to correct these problems.

Though the false bride may have had Falada slaughtered, the instinctive bonds to the mother are not so easily silenced, and are at last activated. The moment the false bride takes away the true bride's last bit of protection once and for all and thus attempts to complete her enslavement, the true bride finally rises to take action—making her deal with the slaughterer. This is the narrative's turning point. Just as the horse heads of the Teutons were supposed to drive away evil, so Falada's head should drive it away. But in my view, what is even more important is the fact that the girl now turns back to Falada—an opportunity that may have been created when the false bride abandoned Falada in favor of the prince.

Translating this scenario into the psychology of people possessed by a lust for power (a compulsion which they feel they should be able to resist but can't), we can say that once they achieve some success, however small it may seem to others, they feel a certain sense of relief that allows formerly obscured dimensions of the personality to emerge. The goose maid goes out every morning through a dark gate to an open meadow, suggesting that she must pass again and again through a dark place in order to come to a clearing where she can find herself. Of course, the dark passage also suggests that this unhappy, anonymous phase of her life is just that—a "passage." What makes this a passage and not merely a dead end is the words she exchanges with Falada, who repeats what the drops of blood had said at the beginning of the folktale, reminding her of the positive mother—both personal and transpersonal.

Now the true bride begins to come to terms with man. She confronts two aspects of the masculine as she first toys with the youthful aspect portrayed by Conrad, and then meets with the fatherly aspect portrayed by the old king.

A young woman who is closely bonded to her mother, and who has a distant or nonexistent relationship with her father, is likely to have a "shadow" relationship with a man her own age, like that of the false bride and the king's son. Until such a woman has struggled with the childlike and the fatherly aspects of the masculine, it will be difficult for her to be real with a man. Many such women have relationships with fatherly men who have a very youthful side, or with very youthful men who will play with them and yet who fail to fulfill their longing for a father. The same thing is true on the subjective level, of a woman's relationship to her inner masculinity. Here the inner, fatherly dimension would most likely feel bound to convention, tradition, and dependability, while the youthful dimension wants to express something new, creative, and evolving.

How does the true bride behave with Conrad? First she shows him her hair. When he is enticed by its beauty and succumbs to the desire to have a few strands for himself, she invokes her magic winds to whisk him away. A woman's hair is closely linked with her erotic power, reminding us of fairies who comb their golden hair in order to seduce and bewitch men. This is exactly what the goose maid does. When Conrad is seized with the desire to play with her, his hat is swept away by her marvelous breezes. It is amazing how powerful the goose maid has suddenly become—now she's practically a witch! Her power is in part derived from her reconstituted relationship to Falada, but also to her instinctive confidence and a degree of personal power. She sends Conrad away so she can put her hair in order—that is, so that she can bring form and control to her erotic fantasies and braid them together. This exasperates Conrad. For two days he controls himself, but soon his patience runs out, and he goes to the old king to complain.

The goose maid's behavior with Conrad represents a kind of "teasing" that enjoys the initial infatuation of an erotic encounter but prevents the real relationship that might have followed from developing. In this behavior, which seduces the boy and sends him away as pleases her, she lets Conrad experience her "feminine power" and her "witchy side." In a more subjective way, this correlates with a situation in which a woman is fascinated by an idea, basks in the enjoyment of it, only to let it slip away without coming to fruition.

At this point in feminine development, the stage is set for the "fatherly" to enter, that side of the masculine that guards against such flightiness. Thus such women, fascinated by the *puer aeternus*, run the risk of opting in the end for law and order, thereby avoiding what is new and trying to emerge, and missing out on the divinities of spring in their lives.

In our tale, a father figure appears who in fact finds out what the goose maid needs. The old king asks her what is troubling her. He wants her to give voice to her pain so that she can set things right for herself. Having passed through this phase of growth by tending geese, she has come to a place where she can be honest about what has happened. Here is a classic therapeutic situation: Until now, she has been unable to divulge her secret. Now she is no longer capable of carrying it alone. The keeping of a secret may be the first fundamental step toward autonomy, especially because one has to tolerate the anxiety that the secrecy produces. But finally the secret must be let out; one must be relieved of its oppressive energy by sharing it with someone else.

It may seem curious that the young woman confesses to an oven rather than the king. But in addition to the necessity of sharing the secret, she must also keep her oath. After all, it was through the oath that she came into a relationship with heaven, and it is through keeping true to the oath that she stays true to herself. And so the goose maid climbs in the oven and begins weeping and bemoaning her fate. For the first time she says to herself, "If my good mother only knew, her heart would surely break in two"—at which point the king bids her to come out. In thinking about how she crawls into the oven in order to make her situation conscious, after being overwhelmed by the shadow, we are strongly reminded of processes of transformation, of death and rebirth, and incubation—of being "baked" to completion.

In the shelter of this extremely protected, womblike place, she can at last put together the pieces of all that has transpired and realize the full extent of it. Giving voice to her secret produces torrents of emotion. Finally, she too can say the little verse about how her mother's heart would break, which previously was recited only by the drops of blood and by Falada—things magically connected to her mother, that represented intuitive hunches more than conscious recognitions of her own terrible situation.

Applied to a woman's psychology, this symbolism suggests that a horrible, anxiety-provoking secret can only be aired when the woman has become capable of taking up a relationship with her masculine side. Given proper and sufficient shelter, she can become clear about what has happened, and be honest with herself about it. Then her true personality can unfold—she can be dressed in her "royal clothing."

That which was producing anxiety behind the scenes can now be seen for what it is—or, as in the tale's language, the chambermaid's plot can be exposed, and she can finally be punished. It is common in folktales for so-called evil figures to conceive of a horrible punishment for someone else that is then imposed on them, usually to their great surprise. The details of such punishments are usually extremely cruel. In some ways, the barrel reminds us of the oven, but the barrel's nails have little to do with the great mother's incubating aspect, and a great deal to do with her rending aspect. With her choice of a method of execution, the chambermaid indicates that she belongs to the phenomenology of the rending great mother, along with the Indian goddess Kali and others. The story tells us that this aspect can now be dispensed with.

Endings such as this one always leave me feeling somewhat dissatisfied. There may be a certain logic about eliminating the problem in this way. But I believe that if we imagine a real person whose positive maternal background and naiveté have gotten her entangled in life, whose negative maternal aspect has thus become constellated and whose power shadow has emerged, particularly in relationships with men, who has enormously expanded her autonomy after tolerating bitter loneliness, and who has developed her differentiated masculine side —the simple disappearance of her shadow is no real solution.

One can of course say that this chambermaid side is so dangerous than it must be split off and rejected. But here I have to ask myself to what extent the psychology of the folktale reflects a particular time and place. In former times, when collective systems of values may have been stronger, destructive tendencies and darkness in general were repudiated in order that people could reach the light. I believe that in our time this is no longer an option; rather, we must attempt to live out the chambermaid aspect of ourselves without falling completely into her power. Clearly, women who are free to take up a relationship with their masculine sides will have less need to assert themselves with unconscious power strategies. Even so, I still do not think it is possible to make these shadow sides disappear completely, for example, by fighting "sadistically" against them. It is more frightening for the chambermaid to be torn to pieces, for then there is the possibility that she will reemerge later as a ghost.

Only once the chambermaid is dead can the young king and the true princess marry. Only then can the relationship to the masculine really be lived; separated from her mother, the young woman gains autonomy. (Concerning the prince's lack of independence much remains to be said!)

As to the subject of anxiety and the ability to cope, we can say that this folktale is concerned with the anxiety that results from being overcome by a negative side of one's own nature, a side that comes up in dealing with separation from the mother and one's relationship to the masculine. For that matter, it concerns the anxiety produced by any separation, when suddenly sides of oneself that had been judged as inferior in one's previous life rise to power and seem to overrule all other dimensions of the personality. The tale suggests that this anxiety must be suffered and tolerated—after all, the chambermaid goes on with her plan, and there is no return to the mother (if there were, this would not be a true folktale). The heroine finds a way to cope with the situation, and through the development of parts of her psyche (the masculine aspects of her nature) that previously lay fallow, the issue that produced anxiety can be consciously formulated—in this case in the form of a relationship to a man.

■ Graycoat

Once there was a king who had three daughters. One day while hunting, he entered a forest and lost his way. On and on he wandered without finding his way out of the forest. As nighttime approached, he met up with a man wrapped up completely in a gray coat so that nothing could be seen of his head or legs. Graycoat asked the king what his destination was, and the king responded that he was lost. Graycoat answered that if the king would agree to give him the first thing that he met when he got home, he would lead him out of the forest. The king agreed, thinking that it would probably be his dog, running out to greet him. Once the king was safely out of the forest, Graycoat said he would come the next morning when the clock struck eight to collect what was his due.

When the king arrived at home, his youngest daughter was the first one to run out to meet him. The king motioned to her to stay where she was, but she only quickened her step, and arriving by his side, threw her arms around him. This, of course, made the king very sad, and prompted his daughter to ask what the matter was. When she heard, she was at first frightened, but then she comforted her father, telling him everything would turn out for the best.

The next morning the youngest daughter put on a black silken dress and prepared for her departure. When the clock struck eight, a coach pulled up, inside of which sat Graycoat for everyone to see. Heartbroken, the king led his daughter to the coach and handed her over to Graycoat, who then drove away with her.

After traveling some distance, the coach approached a huge mountain, which opened itself before the travelers so they could drive inside. Within was a large castle with a great many rooms and a rich array of food and drink. Graycoat gave the princess a set of keys and told her that she could visit any room she liked except for one particular room in the cellar.

Soon thereafter, the curious girl set about investigating the castle from top to bottom. After much searching, she came to the forbidden room and began to wonder what was inside, so she unlocked the door. No sooner had she done this than Graycoat himself came running out at her, frightening her so badly that she was struck speechless. "My child," he asked her, "what did you see when you looked in that room in the cellar?" But he obtained no answer from the speechless girl. "If you cannot answer me, you shall sit in a fig tree where ravens shall tear the flesh from your bones." After this, he removed all her beautiful clothes, leaving her stark naked except for her golden necklace and golden rings. Then he led her out onto the mountain, to a tall fig tree that stood at the top. She must climb into the tree and not come down, he said. The speechless girl did as he had told her.

As it happened, on that very same day the prince of the region had summoned a major hunting expedition. On the hunt, the hounds halted beneath the fig tree in which the princess was sitting and began to bark and bay without ceasing. Hearing the din, the prince told his hunting companion to climb the tree and see what was up there. When the companion reached the top, the princess removed a golden ring from her finger and held it out to him. He took the ring, climbed back down the tree, and told the prince that he could not find anything. But the dogs kept barking around the base of the tree. So the prince sent another hunter to see what he could see. The princess removed a second ring, held it out to him, and he took it. But the tree was so thick with leaves, the hunter couldn't see the princess behind it. Finally the prince climbed up himself. Now the princess took off her golden chain and held it out. But the prince took her hand instead and helped her down from the tree. When he saw that she was naked, he wrapped her in his coat and took her home with him. There she received the finest clothes of silk and satin and was so beautiful that the prince married her. But she was still mute and couldn't say a word.

After a year had passed, the princess gave birth to a son, but three nights later, while she was sleeping, Graycoat came and took the child away. The next morning, the royal parents saw that their baby was missing and found instead an intestine wrapped around a chair, as if as to signal that the child had been murdered. The young mother cried, but no one could tell her what had happened.

One year later, she gave birth to another son, but on the third night after his birth, Graycoat came and took him away, too, and the next morning the princess again found nothing but intestines wrapped around a chair. The old queen demanded to know what sort of creature the prince had found in that tree, who couldn't utter a single word and swallowed her own children. But despite her harsh words, the prince still would not leave her.

When another year had passed, she bore yet a third son. But when Graycoat took this child away, the old queen complained so bitterly that the princess was summoned before the judge. The queen asked her what she had done with her children, but she could say nothing, and only cried. And so it was that she was sentenced to death by beheading.

She went patiently to the executioner, who was standing ready with his large sword. But just as she was about to lay her head on the block, a golden coach drawn by four black horses suddenly pulled up, and a terrible voice called out, "Halt!" Everyone was still as Graycoat stepped out of his coach. He went to the princess and asked her again, "My child, what did you see when you looked into that room in the cellar?" Finally, she found her tongue again and said, "What did I see? I saw an enchanted Graycoat!"

At once Graycoat was transformed, and stood before her as a handsome prince. Then, taking all three children out of the coach, he explained that he was the one who had taken the children and that he was their rightful father; that now he was redeemed, and she was his wife. He led her to his coach and together they drove off to his castle, which now stood on top of the mountain again. The prince became king, and she became the queen.

———————

The foregoing folktale[9] was recorded in Germany, though the figure of Graycoat is more common in Swedish folktales, which unfortunately are no longer accessible. Bolte-Polivka[10] mention a Swedish parallel in which Graycoat—invisible by day— appears to the youngest daughter in a dream, as a beautiful youth who forbids her to open a latch in the floor of his room. When she does so, and sees Graycoat, she is so frightened that she loses her speech and falls down as if dead. When she awakens, she finds herself in a wilderness. The rest of the story is the same as in the German tale.

I find the parallels between these two versions significant because they reveal certain underlying patterns in all folktales of the "animal bridegroom" type. The best-known tales of this genre are probably those of the "The Lilting, Leaping Lark" type. In these, a girl is promised to an animal, usually because the father stole a flower from the animal's garden. Out of love for her father, the girl stays with the animal and, by loving him, transforms him into a prince. In other variants, the animal bridegroom is a gorgeous prince by night and a terrible animal by day. By refusing to abandon the animal, or by burning its hide at the right time, the girl changes the animal into a prince, who then informs her that he had been enchanted by a witch. Often, however, the young heroine burns the hide at the wrong time, and is sent on a seemingly endless quest in search of her prince.

In both the Graycoat and the animal bridegroom types of folktale, the male figure has two sides: one is horrible and frightening, the other beautiful, beaming, and youthful. The disguise can be quite a cause for shock. From the end of the Graycoat tale, we learn that this awful appearance is the result of an enchantment—though we never learn who did it—from which the youngest daughter saves him. In the animal bridegroom folktales, women are not just afraid of the animal; they are disgusted and horrified by him—especially his request to be kissed. Graycoat, however, is more disguised, mysterious, and powerful.

Another parallel between these tales can be found in the father's unwittingly promising his youngest daughter to the animal or monster—or in this case, Graycoat—in order to save his own skin.

Let us examine the father's plight in the Graycoat tale more closely. He sets out for a hunt, but instead of locating game, gets lost and meanders around in the forest, unable to find his way out. Usually in folktales, those who get lost in the forest eventually meet up with the thing or person responsible for their losing their way—in this case, Graycoat, a figure wrapped entirely in a gray coat. That the king has gotten lost in the forest means that he has not succeeded at his attempt to shoot something and thereby bring it out of the instinctual sphere of the unconscious. Instead, he has gotten lost in his fantasies, so that consciousness has lost its orientation and become wrapped up and disguised, so to speak, in fantasies. Behind this loss of orientation stands the veiled figure of Graycoat. It is not clear what is underneath this shrouded figure—all that is known about him is his grayness. Notice that the story does not at first tell us that this is a figure that produces horror. Gray is the color of a lack of desire, a lack of clarity, a fogginess; it lies somewhere between day and night, a dusky ambiguity. Though we can infer that there is something within the gray coat, at the beginning it is simply hidden, unknown. (Notice also that the king meets Graycoat just as night is about to fall.) If we choose to view the folktale as the description of an individual's situation, we might say that here is a father without a woman, who gets lost in his fantasies, cannot find his way out, and therefore loses all desire. He then becomes depressed and is confronted with the ambiguity of his fantasies. At this point, the man in the gray coat tells him that he could liberate himself from this situation if he sacrificed something.

The king is ready to make a sacrifice in order to get out of the forest of his confusion, but he imagines that this will entail nothing more than his dog, while in reality it is his daughter who is at stake. This is a typical folktale motif: a father gets into difficulties, a quick solution presents itself, and he gladly accepts it. It then falls on the child to resolve the underlying problem. (This motif appears not only in folktales, of course, but in daily life as well.) The father's thought that he will probably have to sacrifice his dog may refer to an all-too-human, partial readiness to sacrifice: we know we have to make some sacrifices, but let it not be anything too dear, please. On the other hand, the dog also symbolizes one's instinctual side, which is to a certain degree faithful to consciousness and is often associated with male sexuality. So, perhaps the father thinks he has to sacrifice a bit of his sexuality while in reality he has to sacrifice his daughter, along with his relationship to her. The daughter's joyful reception of her father and her willingness to submit herself to the fate he has gotten her into hints that theirs is an incestuous relationship. Moreover, the mother is not to be found. This quality of incest-in-disguise becomes visible in the figure of Graycoat, who by meeting the king indicates to us that he is an aspect of the king—a disguised, rather than admitted, wish to have his daughter. This is the source of the ambiguity.

The folktale can be seen to illuminate the issues of a daughter who lives in an incestuous relationship with her father. In this murky world, the fantasies of both father and daughter—especially those that are of an erotic and sexual nature—play a large part. The folktale describes how a woman's relationship to a man and to a sexual partner may take shape, given this background. At the same time, we can understand the girl's incestuous bond symbolically, as an excessive attachment to the reigning norms of consciousness embodied by her father, which make it impossible for her to experience either a man or her own masculine sides as "wholly other." In this condition, nothing really new can happen. This is probably why her children are taken away from her again and again; they symbolize newness, and the product of a relationship with something that is "wholly other."

Let me illustrate the "Graycoat" issue by means of a practical case. A woman of thirty—married, with three children—looks for a therapist because her life has become intolerably boring. She is the youngest daughter in a large family. Her mother died when she was ten years old. She describes herself as her father's favorite. She shares her father's political interests and follows him at a young age to meetings. As far back as she can remember, she has had many erotic and sexual fantasies, as well as precocious sexual relationships with men. She idealizes her father and secretly despises all other men, who seem to her like horrible "beasts." This is especially so during sexual intercourse; she thinks of her father and is "turned off." In each case, she held the men responsible for the "failure," putting them down as incompetent, and so forth. She desperately searches for a man with whom she would not have to think of her father. But all the men she meets, she says, are either "angels" or "brutal beaters"—that is, markedly intellectual or obsessed with hard sex. At a young age the analysand married a man with whom she lives in her father's house. The husband is determined to keep the family together. The woman is also very bound to her father intellectually and does not permit herself to think any thoughts that would be unacceptable to him.

Returning to our story, we find Graycoat bringing the king's daughter to his palace, a castle inside of a tall mountain, a very obscure place. That she puts on her black silken dress suggests to me that she feels a certain affiliation with Graycoat; though black is not exactly gray, it is close. Or perhaps the tale wants to indicate that the daughter carries out her father's wishes willingly.

Inside mountains is where recluses and fairies live. If Graycoat's castle is here, he must live in a realm very far from consciousness, in a very magical world. Thus we may conclude that the daughter's relationship to Graycoat consists of highly unconscious fantasies, triggered by her hidden erotic relationship to her father. At the same time, one can also imagine the interior of the mountain as a maternal realm—similar to the forest in which the king got lost, which was also a place suggesting nature in its maternal aspect. Thus we may say that Graycoat, the king, and his daughter are all under the influence of the mother archetype in its protective and imprisoning aspect. This fits with the incestuous situation, in which the family often feels bound to stay together and compelled to suppress all exogamous tendencies. However, in Graycoat's castle there is a room that it is forbidden to enter. We know the motif of the forbidden room from countless folktales, and we also know that it is the very prohibition that stimulates the desire to enter. Many different kinds of persons, and sometimes animals, are to be met within these forbidden rooms. Usually, one finds there the thing that had been most intently banished from consciousness—and thus the thing most urgently required. The repression is what makes the room's content so unbelievably frightening, or numinous, or both.

In our tale, it is not obvious why the girl is so frightened by Graycoat; after all, she had seen him just before, when he picked her up in his coach. At this point I find the Swedish version more convincing: the daughter dreams at night of Graycoat as a handsome prince. The next day she opens a latch that she had been forbidden to open, and sees Graycoat inside. Apparently recognizing him as the same person she has dreamed of, she is so frightened she loses her ability to speak and falls down as if dead. That she has dreamed of the prince shows she is capable of taking the step, in fantasy, from her father to the unknown masculine figure. However, the figure she see when she opens the latch is not handsome, but a disgusting, terrifying Graycoat. Through this man, she encounters both the *fascinosum* and the *tremendum*. In her nighttime consciousness, sexuality is something very beautiful, but her daytime consciousness is still so frightened by the prospect that she loses her speech. The same was true of my father-bound analysand, who once told me that she found sex at night very nice, but when she thought about it during the day, she found "the whole thing" disgusting. Sexuality clearly had two sides for her: a fascinating one and a repellant one—just as Graycoat has two sides in the folktale.

At the end of the folktale, we learn that the king's daughter is so alarmed and frightened because she has seen an "enchanted" Graycoat. Undoubtedly, it is the incestuous relationship between father and daughter that makes this man enchanted to her. To understand this relationship, it is helpful to turn to the animal bridegroom folktales, where the girl is clearly frightened by the bestial side of sexuality. One could also perhaps say that it is the girl's fear of sexuality—along with an inability to accept the man's intellectual side—that turns the man into an animal.

Graycoat's question is an odd one. At some level, he knows as well as she does what she saw. But at the end of the tale it becomes clear that the girl has to make clear in her own mind that it was an enchanted Graycoat that she saw—not the man's true form, but a man put into a terrible form by means of a curse. The girl feels nothing but unmitigated fear in this situation. She has discovered something about sexuality and relationships with men that has caused her the greatest possible anxiety. She can utter no word, her throat is shut tight, and she remains speechless for some time.

Everything bestial is unnerving. People who behave like animals are profoundly disturbing, causing unbelievably aggressive, destructive, and sexual powers to erupt. For a girl, the experience of male sexuality, of sexuality in general, can be very alarming. But when the situation is complicated by a highly incestuous, secretive relationship with the father, what emerges will be even more alarming. Add to that the fact that the father has not coped with the problem himself and instead denies the incestuous relationship.

The result of all this anxiety is that the girl is led out of the underworldly castle and stripped naked. This brings to mind Adam and Eve in Paradise. Being naked means being stripped of one's disguise and one's protection; one who has been stripped naked is vulnerable, but also ready for a new beginning.

This factor, along with many others, suggests that an initiation is taking place: The girl goes into the mountain, where she experiences something intensely frightening. She reemerges and is made to remove her clothes and climb into a tree. This last step reminds us of the motif of tree-birth—growing out of a tree which embodies the archetypal father and mother in one, a rebirth that has nothing to do with the personal parents. In all of these ways, the girl is being initiated into her womanly being by enduring in her fantasy the anxiety associated with masculinity, sexuality, and her erotic desires for her father.

Such an initiation might take place in the life history of a girl who had an incestuous bond to her father and a great many fantasies about partnership and sexuality. In her most beautiful fantasies, she suddenly realizes that she is terribly frightened of men, who hide their true natures from her, leaving her to experience a shady, murky ambiguity. These fantasies, this shock, bring the girl to a new level of development, the level of a woman.

In the case of my analysand, this phase expressed itself as follows: After working on her material for a while, she was suddenly alarmed to realize that she had actually "wanted" her father, and that behind her rejection of "animalistic sexuality" lay a fear of her own sexual desires towards her father. This anxiety about her own sexuality had reversed itself, causing her to "disparage" masculine sexuality.

In the folktale, once the girl has reached a new level of development, the prince can succeed at his hunting where the old king had been unsuccessful. Because he understands her as a human being, he is able to help her down from the tree. (Those who only wanted her rings did not understand her as a human being).

The fig tree is a symbol of fertility and excess; it was sacred to Dionysus, and thus is a symbol of eroticism and sexuality as well. Like a ripe piece of fruit, the girl is plucked from this tree of eroticism and sexuality, given new clothing and made the prince's wife. But still she is not capable of speech.

As long as she remains silent, the only instrument she has to use in shaping her relationship with her husband is her body; the entire dimension of human togetherness shaped with speech and language is missing. Indeed, at this point in the tale the girl is portrayed as a creature of nature—born out of the fig tree. After this, the prince's mother asks him what kind of a creature he took down out of that tree that doesn't know how to talk. Thus the woman is one whose entire social contact is restricted to her husband, and who may for this reason be terribly lonely; anxiety creates loneliness. And so, despite her attainment of a new level of development, despite her transformation from girl to woman—a woman who is the fruit of a fig tree, where sexuality and eros are given new meaning—she is still mute from the shock of seeing Graycoat's true identity.

This is to say that, even though she has a relationship with a new man who does not resemble him in the least, the young woman is not finished with Graycoat yet. If Graycoat stands for the distorted image the young woman has of men, owing to her fear of her own incestuous desires, as long as he is in the picture, the relationship with her husband will be marked by incestuous wishes and anxieties as well. Whatever she creates with the earthly man cannot exist—hence Graycoat's repeated stealing of her children. As I said earlier, the symbolic meaning of an incestuous relationship lies in its ban on anything new that might emerge.

The child is a symbol for what is new. This is why Graycoat must keep taking the children, stealing them away at night. In all such tales in which children are taken away from their mothers in a more or less violent way, the father is either absent or sleeping. This suggests that the male kidnapper can be seen as a "nocturnal" aspect of the man. Previously, Graycoat was dreadful in the day, now he is dreadful at night. This means that the anxiety-provoking facet of the relationship to the man has become unconscious—but no less powerful. I also wonder if this dreadful, kidnapping aspect might not also represent the woman's suspicion that her husband does not want her to have children. If this is so, Graycoat would then also stand for a very demanding, aggressive male sexuality that tolerates no children. Men who cannot cope with their wives becoming mothers are often under a heavy curse; usually they are tightly bound by a severe mother complex. Thus, when their wives become mothers, their fear of incest becomes intense. In this light, I would have to say that not only does the woman in our story have an incestuous relationship to her father, but the man has an incestuous relationship to his mother. And indeed we see that the old queen, the man's mother, makes sure that her daughter-in-law is brought before the judge. In this way, she hopes to eliminate her.

The young mother suffers, cries, and yet says nothing. One can imagine her suffering: Someone has stolen her newborn baby, which embodies her new beginning and her feelings of deep attachment, having brought it into the world with the pain of her own loins. What's more, now she is accused of murdering the child. Is it any wonder she says nothing? Indeed, one wonders if she even knows what has happened. Graycoat came at night, while she too slept.

The disappearance at night can also be interpreted. Life goes on, new impulses for living are born to the young woman, but these disappear overnight; in the morning she is left only with a feeling of loss and perhaps even with the guilt of a murderess. The new development that had promised to enrich her life has vanished already, and it is probably her fault, she thinks. But the real culprit is this Graycoat problem that has not yet been resolved, and about which she dares not speak.

We say that the young woman is not able to speak; actually, it would be truer to say that she does not want to speak at the wrong time, for, when the time is ripe, she does speak. Her silence has another function as well; it separates her from every trace of the Graycoat experience. Since she left Graycoat, she has bound herself to the positive aspect of the masculine, here expressed in her relationship to the prince. In a real woman's fate, this might appear in a scenario like the following. Through the hidden incestuous relationship to her father she develops a horrendous fear of the "terrible" aspect of sexuality—of forbidden sexuality. There is a huge secret, but she does not admit that it is what is causing her anxiety. And so she sees only the light side of her husband, not the dark, instinctual, aggressive forces that are at work, and that are still obscured by the incest fantasy.

After my analysand had become conscious of her fear of sexuality, and of her incest fantasies, she was gripped by incredible shame about her own "instinctuality." She was now able to give up her "pseudo-life" with her father. At the same time, she tried to steer clear of any situation in which she might be tempted sexually, for fear of the anxiety that might break out. With the sacrifice of her incestuous fantasies, her political interests took on a life of their own. As in the folktale, she went through a phase of denying her fascination with the world of Graycoat.

The Graycoat world and its incestuous interweavings can also be understood symbolically. This gray man in a castle under the earth may represent an animus figure. He is so uncanny and fascinating that there is a real danger the woman would simply decide to eke out an existence with him down there—a decision which, for a real woman, could well amount to psychosis. Such dangerous, seductive facets of her own personality, such fascinating fantasies, are capable of dissolving the ego and therefore must be avoided. They must be denied to the point of self-sacrifice—portrayed in the folktale in the image of the death sentence.

Just as she is about to be executed, Graycoat drives up. Now that which has long been disguised begins to unveil itself. His wagon is of gold, his horses are black. The color gold suggests qualities of the sun—indestructible value and perfection—and is also a symbol of consciousness and all the lighter facets of the human being. The four black horses, on the other hand, are creatures of the underworld. If gold is associated with the spirit, then black has to do with dark, animal drives, death, and rebirth. Graycoat's terrible voice is now more powerful and anxiety-provoking than ever.

Here again it is instructive to recall the tale's Swedish parallel, in which the girl dreams of Graycoat as a handsome young man while Graycoat himself hides beneath the floorboards of her house. There, too, what is required is to overcome the split in one's image of a single man. The carriage that pulls up in the German tale promises to bring movement back into the woman's life and, with its gold and black elements, indicates

that the split in the image of man has finally been overcome—
or now may be.

The way Graycoat asks his question shows just how much of
a father he is to the girl: "My child," he says, "what did you see
when you looked in the cellar?" Closely bound to Graycoat, the
woman in our folktale has in some respect remained a child.
Only when she finds her tongue—and thus overcomes her fear
of Graycoat—can she say what she saw. She saw Graycoat, not
simply a man wrapped in a gray coat, but rather a man under a
curse, a man who provoked anxiety in all who beheld him.

My own analysand's final confrontation with the "world of
Graycoat" took place after she dreamed of her father as Pan,
with the feet of a goat and an erect phallus. The dream so startled her that she woke up. Now she began to understand that
lying beneath the relationship to her father was the problem of
a repressed animal sexuality that was pressing for recognition,
and that this had indeed long been her father's problem. When
the princess of the folktale speaks to the enchanted Graycoat, he
turns into a handsome prince, brings the children back, and
announces that she is his wife. In the parallel Swedish tale,
Graycoat falls into the ashes, out of which rises a beautiful
prince. The castle is no longer enclosed within the mountain,
but stands on top of it, symbolically liberated from its disguise,
its need to hide. If we interpret enclosure within the mountain
as confinement in the maternal realm, we can say the man has
been redeemed from his complex by his wife's love. In folktales
of the animal bridegroom type, it is usually some witch or other
woman of the forest who has transformed the man into an animal. We can interpret this in two ways: In order not to lose
touch herself with the animal aspect of masculine sexuality, the
mother may expose her daughter to it so intensely that the girl
has the feeling that men are little more than "dogs" than human
beings. Conversely, a son who is still basically in love with his
mother will show women primarily his sexuality and aggres-

sivity. In this folktale, the mother's influence is a bit more subtle, since a mother figure does not appear until late in the narrative. And yet, we can imagine her behind the scenes, working her influence.

The folktale's conclusion shows the woman's fear of the masculine coming to an end. Once she has overcome the split in her experience of the masculine, the lust awakened by the incestuous relationship with her father is no longer seen as something dangerous. Thus, the folktale also sketches out a course of development from an incestuous to a real relationship with the masculine. The anxiety described in the tale stemmed from both a fear of the instincts and fear of the power of masculinity, which in an incestuous relationship is both fascinating and dreadful.

The story also shows a strategy for coping successfully with that anxiety, in the girl's steadfast refusal to relinquish the positive side of the masculine and in her refusal to speak before it is time, even though it is clear that silence is no solution either. At a certain point in the story, her life comes to a standstill. Her children have all been taken away—they have not been allowed to live, indeed, may not have wanted to live (the deathly boredom of my analysand). In this moment, the old problem arises again. But this time, the woman resolves the problem by giving it a name at the right time. Then all the life-energy that had been missing can return again.

In folktales like this one, in which the portrayal of great anxiety creates a great deal of suspense, a character's ability to tolerate the anxiety almost invariably causes something wonderful to take place. The anxiety vanishes and makes room for renewed liveliness. But it is important to find the right moment for burning the animal skin, for naming the terrible truth. If the naming is done prematurely, the integration and transformation of the anxiety-provoking content may have to be postponed for a long time. We see this often in therapy, where it is crucial that the therapist not address an anxiety-producing secret too early. However, in therapy as in folktales, it is also possible to wait too long to address such a secret. For then the enchanted men—both literal and figurative—remain entrapped in their animal shapes.

■ Nixie in the Pond

There was once a miller who lived happily with his wife. They had money and property, and their wealth increased from year to year. But misfortune comes overnight. Just as their riches had grown, so they dwindled away year after year, and in the end the miller could hardly even call the mill in which he sat his own. Beset by troubles, he found no peace even when he lay down at the end of a day's work, tossing and turning in his bed.

One morning he got up before daybreak and went outside, prompted by a feeling that perhaps today his luck would change. As he walked across the dam, the first ray of sunlight broke over the horizon and he heard a sound coming from the pond. Turning around, he saw a beautiful woman rising slowly out of the water. With her tender hands, she held her long hair over her shoulders so that it draped down over both sides of her white body, veiling her. He recognized her as the nixie of the pond, and in his fear he did not know whether he should run away or remain standing where he was. But the nixie decided to let him hear her soft voice, and so called his name and asked why he was so sad. At first the miller was struck dumb, but when he heard her speaking in such a friendly tone, he was heartened and told her that until now he had lived happily and well, but that of late he had become so poor that he didn't know what to do. "Hush," answered the nixie. "I will make you richer and happier than you have ever been in your life. You must only promise to give me that in your house which has just become young."

"What can that be but a new puppy or kitten?" thought the Miller, and so he agreed to her condition. Then the nixie dove back into the water and the miller hurried back to his mill, feeling happy and relieved.

As he was approaching his house, the maid stepped out of the front door, calling out that he had cause to celebrate—his wife had just given birth to a baby boy. The miller felt as if he had been struck by lighting; he realized that the treacherous nixie had known about the birth and had cheated him. With his head bent low, he went to his wife's bed. When she asked him, "Why are you not happy about the beautiful boy?" he told her what had happened to him, and the promise that he had made to the nixie. "What good is happiness and wealth, if it means I have to lose my child," he said. "But what can I do?" Neither his wife nor the relatives who had come to wish the newborn well had any advice to give him.

In the ensuing months, luck and prosperity found its way back into the miller's house. Whatever he undertook, he succeeded at. It was as if his coffers filled themselves of their own accord and the money in his closet increased overnight. In a very short while, his wealth was greater than it had ever been before. Yet he simply could not enjoy it. His heart was tormented by the agreement he had made with the nixie. Every time he visited the pond, he was afraid she would come up and demand what he had promised her. Never would he leave the boy anywhere near the water. "Watch out," he would tell him, "if you touch the water, a hand will come out, grab you, and pull you under." But as year after year passed without any sign of the nixie, the miller began to feel more at ease.

When the boy became a young man, he apprenticed himself to a hunter. After mastering the skills of the trade, he was taken into the service of the lord of the village. Now in that village there was a beautiful and virtuous girl whom the young hunter grew quite fond of. When his lord noticed the lad's interest, he gave him a small house. Soon thereafter, the couple were married and lived peacefully and happily in the little house, loving each other dearly.

One day the hunter was out tracking a deer when the animal left the forest and turned into an open field. Chasing after him, the hunter laid him flat with a single shot. But the chase had so preoccupied him that he gave no thought to the fact that he had wandered quite near to the dangerous pond. After skinning the animal, he went to the pond's edge to wash the blood off his hands. But as soon as he put his hands in the water, the nixie rose up smiling and put her wet arms around the young hunter, pulling him under so fast that in an instant, the waves had closed over his head.

When evening came and the hunter had not arrived home, his wife began to worry, and went out to look for him. He had told her many times that he had to be careful of the nixie's snares, that he dared not go anywhere near the pond, and so she quickly guessed what had happened. Hurrying to the water, she found his hunting bag lying on the shore. Moaning and wringing her hands, she called her love by name, but in vain. Running to the other side of the pond, she called for him again. She swore at the nixie, but there was no answer. The water's mirrory surface was undisturbed with only the moon's half-face looking sternly back up at her.

The poor woman could not leave the pond. Unable to find either peace or comfort, she paced quickly around its edge, circling it again and again, sometimes quietly, sometimes screaming out loud, sometimes whimpering meekly. When her energies were finally spent, she sank to the ground and fell into a deep sleep. Soon she was overcome with a dream:

Filled with fear, she climbed up between two huge rock walls. Thornbushes and creeping vines clung to her feet, the rain beat down on her face, and the wind ruffled her long hair. When she reached the top of the hill, everything changed. The sky was blue, the air was pleasant, the ground sloped gently downward, and in the midst of a green meadow strewn with bright-colored flowers she saw a tidy little hut. She opened the door, and there sat a white-haired old crone, who gave her a friendly smile.

With that, the poor woman woke up. Finding that the day had already dawned, she decided not to waste another minute but to do exactly as she had done in her dream. The climb up the mountain was indeed strenuous, and everything she saw looked exactly as it had in her dream. The old crone welcomed her kindly and showed her a chair to sit in. "Only some misfortune," she said, "can have brought you to my lonely hut." In tears the hunter's wife related what had befallen her dear husband. "Dry your tears," said the old crone, "I will help you. Here is a golden comb. Wait until the full moon rises, then go to the pond, sit down by the edge, and comb your long, black hair. When you are finished, put the comb down on the shore, and watch what happens."

And so the hunter's wife returned to her home and waited for the moon to grow full. Finally the glowing disk appeared in the sky. Going out to the pond, she sat down and combed her long, black hair with the golden comb. When she was done, she laid it down at the water's edge. Not long after, the water rose up from the deep and a wave broke onto the shore, whisking the comb away as it receded. As soon as the comb had sunken to the bottom of the pond, the glass-smooth water divided itself and rising up came the hunter's head. He said nothing, but looked at his wife with imploring eyes. Just then a second wave came up and covered his head again. Everything disappeared, the pond lay as calm as before, with the face of the full moon shining on it.

Dejected, the woman returned home, but that night she dreamed again of the crone's hut. And so the next morning she went to see the wise old woman. Sobbing, she told her what had happened at the pond. This time, the old crone gave her a golden flute, instructing her, "Wait until the full moon comes again, take this flute, sit on the shore, and play a pretty song. When you are finished, lay the flute in the sand and watch what happens."

The woman did as the crone instructed. As soon as she lay the flute down in the sand, the water rose up again and a wave broke on the shore, drawing the flute away with it. Then the water parted as it had before, but this time not only the hunter's head but the whole upper half of his body rose up. He stretched out his arms longingly toward his wife, but a second wave swelled up, covered him, and pulled him down again.

"Oh, what good is it if I only get a glimpse of my love only to lose him again," said the unhappy woman. Grief filled her heart afresh, but that night her dream led her yet a third time to the house of the old crone. When she went the next day to visit the wise woman, she was given a golden spinning wheel, and told, "Fear not. Our work is not quite complete. Wait until the moon is full, then take this spinning wheel, sit on the shore, and spin the spool until it is full. When you are finished, put the spinning wheel close to the water and watch what happens."

The hunter's wife followed these directions in every detail. On the night of the full moon, she carried the golden spinning wheel to the shore and spun eagerly until the flax was finished and the spool filled with yarn. As soon as she had set the wheel on the shore, the deep bubbled and foamed louder than ever, and a mighty wave came and splashed the wheel away. Just then the hunter's head appeared in a jet of water and continued rising until the whole body was clear out of the water. Quickly the hunter jumped onto the shore, grabbed his wife by the hand, and ran away. But they hadn't gotten very far when the entire pond rose up with a terrific roar, and heaved itself torrentially onto the broad meadow. The fugitives saw their deaths before them. In her fright, the woman called on the help of the old crone, and in an instant both husband and wife were transformed—she into a toad, he into a frog. Though the flood drenched them, it could not kill them. Still, it pulled them apart and carried them off separately.

When the water had subsided and both were again touching dry ground, they were returned to their human shapes. But neither one could find the other; they found themselves among strange people who had never even heard of their homeland. High mountains and deep valleys lay between them. In order to make a living, each tended sheep. For many long years they drove their herds through distant fields and woods, their hearts full of grief and longing.

One day, when spring rose out of the earth once again, both went out with their flocks, and as chance would have it, they happened to run into each other. From a distant mountain cliff the hunter spied the other flock and drove his sheep downward to where they were. Both entered the valley at the same time without recognizing the other. Yet they were pleased not to be alone any longer. From that day forward, they drove their flocks together. Though they said very little, they still felt comforted by the other's presence.

One night when the moon was full and the sheep were at rest, the shepherd took a flute out of his bag. He played a beautiful but sad song, and when he was finished, noticed that the shepherdess was crying bitter tears. "Why are you crying?" he asked. "Oh," she answered, "the full moon shone exactly like this the last time I played that song on the flute, when the head of my beloved rose out of the water." As he looked at her, it was as if blinders fell away from his eyes, and he recognized his most beloved wife. Seeing him with the moon shining in his face, she recognized her long-sought husband as well. They fell into each other's arms and their lips touched. We needn't ask if they were in bliss.

This folktale[12] opens with the description of a miller who has been reduced to utter poverty. Full of worries, he does not know what to do. We are told that the miller used to be very rich and lived very well.

A miller earns his wealth—or rather *earned* his wealth—by means of the coursing water that drives the wheels of his mills. Thus we might say that the "miller" represents an attitude of consciousness that knows how to get something out of the energies of the unconscious; he uses it to grind grain for making bread, for providing nourishment. But in the case of our poor, destitute miller, this possibility suddenly seems to have vanished. The tale does not let us in on exactly what has gone wrong, but one possibility is that the river under his mill has dried up. In other words, the river of the unconscious has dried up—and with it, the miller's life as well. Such a dried-up life is depicted in this and other folktales as a state of poverty, as it is often depicted as a state of childlessness. Here, we are not told whether the miller has had other children, but we do know that the birth of a son is announced as if this were something that has not taken place for some time.

If in fact the miller and his wife are childless at the story's beginning, we could say that theirs is a relationship from which the life has been drained. Because the miller in this folktale is the figure who carries the greatest sorrow—at least to begin with—we may suppose that it is the masculine frame of mind that has lost its relationship to the feminine. More immediately, it has lost its relationship to the emotions, to eros. It has also lost its understanding that things must take their own course—poverty following on wealth, wealth again on poverty. That life moves in cycles.

One morning, when the miller is without a clue as to how to proceed, he meets a nixie in a pond. In his fear, he does not know whether he should stand still or run away. But with her soft voice, the nixie seduces him into telling her his woes, and promises to help him under the condition that he give her whatever has become "young at home." Thus he is rescued from his plight, but leaves whatever has new life at his home to face the nixie.

Here we have the same motif as found in "Graycoat," though here it is not the daughter but the son who is being given away. The fact that the miller doesn't know that a baby has been born in his own house is yet another indication of how inattentive he has become in his relationship with his wife, and how stubbornly pessimistic he is. At the very moment he thinks his poverty is complete, a child is on its way.

Turning now to the nixie, we must ask, what does she signify? Mermaids are extremely seductive female figures who lure men down into the water—just as in our folktale (and in Goethe's famous poem, "The Fisher": "She drew him in partly, he sank in partly, and nevermore was seen. . ."). They seem to want so badly to have a man because they themselves have no soul. They are supposed to be unusually passionate and always manage to make a man lose his head, so that he abandons himself completely to his passions, emotions, and fantasies. The danger is that the man is either drawn into the unconscious, or, if he loses his nixie, he is plagued by wild longings that drive him to leave everything behind and wander off into unreality. Sinking into the water, getting lost in the woods, losing one's way in the desert—these are all symbols for the experience of being overwhelmed by the unconscious. And yet I think that there is an important distinction to be made between losing one's way in the woods and being pulled into the water. Those who are lost in the woods at least still have the ground to stand on; water is more immediately engulfing. Thus I think that being devoured by a nixie indicates a more severe regression into the unconscious than a mere getting lost in the woods, as happened to the king in "Graycoat."

The encounter with the nixie portrays a man's intense anxiety about, as well as his longing for, passionate, nature-bound emotionality. It was just this fascinating, dangerous emotionality that had no place in the miller's marriage. And that is why it is now so seductive to him.

Behind the nixie, of course, is a mother goddess—especially in her aspect of a love goddess, such as Aphrodite or Venus. This explains the nixie's numinosity. She is not "merely" a nixie; she possesses divine power. Vestiges of such nixie figures have survived into contemporary life—for example, in the common fear of seaweed, which is often associated with a dread of nixies who might pull one to the bottom of the sea.

Of course, it is not only a question of fear, but of the *fascinosum*. In folktales where human beings have allied themselves with nixies, or in which men fall in love with mermaids, the human is always longing for something completely different: "depth," transcendence, or a going beyond the bounds, to an unknown realm that is both frightening and fascinating. The depth and transcendence is usually sought, and found, in eros and sexuality. He who gets involved with nixies should be prepared to be devoured and taken away. Going beyond the bounds can lead to dissolution. No one who is seized by these emotions can escape being changed by them.

In our tale, however, the miller is not required to pay the price himself. In his highly restrictive situation, the nixie appears to him as one who brings new hope. Although he is enticed into telling her his woes, he is not really "seized" in any profound way. Speaking psychologically, we could compare this with a situation in which one feels "dried up," nervous, and depressed. Suddenly one is captured by an uplifting emotion, but there follows no real surrender because the emotion seems too frightening. It is enough to have a taste of that new and exciting (and dangerous) something. But, in order not to dry out again, one must allow oneself to be more profoundly seized. For the miller, this first, glancing encounter with the nixie has been enough to uplift him—albeit temporarily—as can be seen through the immediate restoration of his wealth.

When nixies appear as water-women who take something away from a human being, it is usually because they have been taken too lightly and have not been allowed to participate enough in life. This explains why they are threatening. When men, or entire collective systems, split off their feelings from consciousness, the longing grows—as does the danger of being flooded by emotions. An "addiction to emotion" then sets in.

The miller's entire future—and his family's—is now over-shadowed by the anxiety that the nixie could collect on her promise and come to take the boy away at any time. That the anxiety is shared by all is shown when the father speaks about the problem to the relatives who come to wish the boy luck.

What is it, then, that threatens the boy? What is the problem that he must resolve? I believe that folktales in which a child is "sold" or "promised" to someone else indicate problems that the parents have failed to resolve and that therefore fall on their children to handle. Here we see the son becoming smitten by love, gripped by emotions that take him far away from human life, and a longing that plunges his daily life into a maelstrom. Like the animal bridegrooms, he, too, will become enchanted, but he won't be there as an animal for his wife to touch—he won't be there at all.

As I see it, enchantment by the nixie expresses an attitude to the feminine, and to the great mother, that has to change. The nixie dwells in a pond. A pond is also where Mother Holle, a major mythological figure, launches children into life and takes them out again. Thus, pond-matters often have to do either with creativity and life, or destruction and death.

A sense of fear, of imminent danger, suffuses the entire folk-tale; but the tale also shows us various ways of dealing with this fear. To begin with, the father tells everyone about his problem.

This is probably a technique for soothing anxiety—which is always less intense when shared among a group than it is when carried by an individual.

Another measure the father takes against anxiety is to warn the boy not to go near the pond lest the water-woman pull him under. This works quite well in subverting the danger—at least until the boy grows up. However, it seems that the nixie is not interested in him until he has grown up—that is, until the issues of love and sexuality become relevant. At that point, he apprentices himself to a hunter—another strategy for avoiding anxiety, and a very clever one indeed. For now he has regular contact with animals of the forest and the air, but not with those of the water. In other words, he becomes familiar with the realm of the vegetative unconscious and also develops a very conscious and goal-oriented attitude, but he keeps a safe distance from the dangerous pond.

Again we can find a parallel to this situation in therapy. Sometimes we encounter persons for whom a certain dimension of the unconscious may be very dangerous and overpowering. They then cultivate other dimensions of the unconscious and thus gain in ego strength, so that eventually they may be able to go on and confront what is threatening them.

Having learned his trade and joined the services of the village lord, he now grows fond of a girl whom he takes for his wife. Also, the lord he serves gives him a house. Thus we see that the hunter has become capable of a relationship to a woman, to the feminine—even if it is a rather modest relationship (the house is small). They live together in peace and happiness and love each other dearly. It is at this point that the man becomes interesting to the nixie—the moment in which he loves someone, his emotions break out, his passion is released. If he was ever in danger of being devoured, it is now.

In folktales, we often find deer luring heroes into distant regions in which they meet something or someone that causes a

transformation to take place. A deer in flight can express a longing for something for which no words can be found, a sense of
one's being pulled toward something unknown. In this tale, the
hunter kills the deer, and thus we might think that he kills his
longing. And yet, because he has to wash its blood off his hands,
we can say that the deer has led him to the nixie and to the pond.

When he does not come home, his wife suspects something
because she knows he had always been on guard against the
nixie's snares. When she goes to the pond and finds his hunting
bag, she knows instantly what has happened. She calls to him,
but he gives no answer. She has now lost all contact with her
husband, and cannot find a way to reach him. From the *Odyssey*
we know that those who hear the sirens go mad. Though we
cannot say whether the hunter is psychotic or depressive—or
has simply stepped into the trap of a real nixie—we do know
that the woman can no longer talk with him. In effect, the relationship is over, his longing for her has ceased.

This image illustrates what can happen to a man who has
long kept in check his infatuation with nixies. It also shows us
the *fascinosum* and danger of an unconscious content that is
bound up with emotionality, the body, and love. The man may
have been able to protect himself from this fate by cultivating
strategies of avoidance, but at some point the object of anxiety
would have to be dealt with. For avoidance cannot completely
erase longing, and longing leads us to the things that we are
afraid of. Particularly in the province of the nixies, the thing
that triggers anxiety can neither be recognized as a projection
nor be approached through rational inquiry. Thus, one is simply overpowered and, having fallen into the trap, no longer visible to the world, no longer present or comprehendable.

A twenty-four-year-old man I know—a very rational, collected, and capable student of economics—has a "nixie-anima"
of this variety. In reality the man suffers from an intense fear of
"nixie" women. They destroy his calm and poise and flood him

with sexual and erotic desires that throw him into a state of confusion whether or not he acts upon them. His girlfriend is not a nixie. But whenever a nixie comes up in a dream, he becomes submerged in erotic and sexual fantasies. For days on end, he shuts himself up in his apartment, which he darkens with green curtains, and celebrates these fantasies. During these episodes, he is unavailable to his girlfriend. No matter how desperate she may be, she cannot get through to him.

In the folktale, the center of gravity now shifts to the woman, and thus my interpretation will now shift to her perspective. Having been abandoned, she is desperate, and will not leave the place of her misfortune. Circling around and around the pond, she gives voice to all her grief about the separation, mulling over her tragedy from all angles. Finally, she sinks to the ground exhausted, falls asleep, and has a dream. Though all of her conscious efforts, moaning, and wailing have been of no use in bringing her man back, they probably do bring on the dream.

The arduous climb up the mountain in the woman's dream can be interpreted as a journey to "get over" the negative aspect of Mother Nature, who is in some respects comparable with the nixie. Discovering a wise old woman who is ready to help, the woman encounters an embodiment of the good aspect of the mother. This aspect of the mother archetype has nothing to do with the water, but in her trim little house in the green meadow strewn with flowers in bloom, is definitely a part of nature.

The dream foretells a feeling of liberation that will come after a phase of hindrance. That is to say, once the hunter's wife has overcome certain difficulties, she will have the sense of being helped. The dream promises that she will not always have to do everything herself and leads her to expect that the situation will clarify itself. In other words, the anxiety will abate. Here the folktale makes no distinction between dream and reality. Whatever is seen in the dream is viewed as a direct suggestion about concrete actions that should be taken in life.

The woman's arduous climb up the mountain can be compared with the hunter's laborious emergence from the rising tide, which comes later in the story. In my view, it suggests that by engaging her own possibilities or working on herself, the woman can resolve the problem of being devoured by the dark aspect of nature. It is easier for her than for her husband, since the pond has not swept her away. But, to the same extent to which she can work on the problem within herself, she redeems her husband as well. If we see the folktale as the description of a dynamic within a couple, we may assume that both partners have the same basic conflict—that of being threatened by an entrancing, devouring, proliferating nature-sprite who is capable of pulling them away from life. If the nixie represents such a dangerous tendency within the man, then we can assume he has married a woman who has this nixielike, nature-sprite quality, but who probably represses it completely.

To illuminate with a case study: A married man and woman, each about forty years old, sought therapy. The man told of an overpowering fascination for nixie women, who with each experience pulled him further away from reality. His encounters had little to do with relationship, but were moments of a sinking, orgiastic experience. The woman he married seemed the exact opposite of a nixie: she was even-tempered, dependable, not in the least conscious of her power to seduce, "there for him," empathic. One could even say she was "anti-nixie"—so very antinixie that one was obliged to ask certain questions.

One day the man fell helplessly in love with yet another nixie woman. It was terrible for him, and no less terrible for his wife. When they came for therapy, not long after this happened, they asked what they could do about their situation. Looking into her dreams, the woman was very quickly confronted with her own nixielike sides. To her great surprise, she had no difficulty in developing these facets of her personality. The couple resolved their marital issue after each worked intensively on the development of their undercultivated sides.

Returning to the folktale, the hunter's wife's ascent up the mountain to find the woman with the white hair strikes me as an ascent to the "light" feminine. The place where this old woman lives gives the impression of orderliness, and the work that lies before the hunter's wife has to do with making order.

The old crone is very empathic, and it is to her that one comes after having suffered some misfortune. She is a kind of mother—archetypal rather than personal—whom one seeks out for advice. At the level of the individual psyche, this figure would suggest that the hunter's wife is approaching a deeper level of understanding. Initially, she may not have grasped what was happening to her. Once she does, she is faced with the absolute finality of her tragic loss, and with her inability to do anything about it by her herself. Then, after going through a period of grief work, a dream, an intuition, or an encounter gives her an idea about what can be done.

In other words, after the negative aspect of an archetype—in this case the nixie in her purely devouring aspect—has done its work and its effects on the psyche have been emotionally integrated, the positive aspect of the same archetype is constellated, pointing the way toward growth out of the situation. Of course, such a process requires a quality of emotional openness such as the woman in the folktale displays. This is someone who is truly open to intuition, dreams, and the irrational in a positive sense.

It is a basic experience and conviction of Jungian psychology that if one recognizes and suffers through a present difficulty in the psyche, its complement will begin to constellate itself; thus, if one does not avoid the negative, the positive aspect will make itself known. When a woman in such a situation has dreams like those of the hunter's wife, we would encourage her to use active imagination to contact the wise old woman (a wise, feminine aspect of her own psyche), who knows how to proceed.

One would encourage the woman to seek the advice of this "inner" woman.

Clearly, the woman in this folktale has fewer problems dealing with her own nixie nature than her husband does, and this is what allows her to make the required developmental step. The first gift she receives is a golden comb, which she is instructed to use to comb her long, black hair in the light of the full moon. The comb is an erotic instrument that nixies use to entice men with, by combing out their long golden or black hair. By combing her hair in the full moon, the hunter's wife consciously takes over a role the nixie has played for her unconsciously: one could say that she becomes conscious of her own powers of erotic seduction, thus deliberately becoming a bit of a nixie herself. And indeed, this act succeeds in bringing the hunter's head up out of the rising tidewaters—i.e., it causes him to begin reemerging as a person. Also, in addition to its erotic connotations, combing has the meaning of bringing order into the hair.

The man looks imploringly at his wife and then sinks back under the waves, leaving his wife to quit the pond in dejection. But again her dream brings up the image of the old crone. The woman must once more wait until the full moon, play a song on the flute, and then put the flute down again. The flute is another instrument nixies use to seduce men. Its soft, lilting tones awaken a man's longings, emotions, and feelings. Flute tones awaken something "transcendental" in men as well—a longing for the eternal. Thus we might say that by playing the song, the hunter's wife seduces her husband back from the realm of the nixies. Now the upper half of his body becomes visible, but grief fills her heart all over again when he sinks back down below the surf. What is the point in seeing him again and again without being able to rejoin him?

At the next full moon the wife is to occupy herself with a spinning wheel, by spinning a spool of yarn by the water. Spinning is something that the great mother goddesses do; they spin the threads of fate. Clearly this is no activity for a nixie. Indeed, one has the feeling that everything about the nixies is chaotic and in disarray. Spinning means making order out of chaos, making a thread to follow that would lead one out of this chaotic, emotional, instinctual situation. But spinning also suggests fantasizing: we say that someone is "spinning out a fantasy," or "spinning a yarn." Work of a methodical nature encourages us to spin out fantasies. It could be that, when the woman spins by the water in the full moon, she fantasizes about her life with her man and thus begins to see him again for the man he is. Fantasizing about someone can often help us to see them in perspective. Sometimes we need to fill a relationship with our positive projections before we can overcome what is destructive about it. By believing something positive about someone, we arrive at the faith that a man can free himself of his fascination by a nixie, which can give him an added impetus to accomplish the task and, ultimately, change dramatically the dynamic of the couple. Of course this is easier said than done!

In the folktale, at least, the wife's fantasies do help. The man arises out of the water a third time, and they flee with each other. The fact that they both must flee means that danger is still present. Indeed, the pond immediately floods over, and a deep regression takes place. In the very moment that they have gotten through the disembodied phase of their relationship, the water washes over them again, now bringing both of them into the realm of the nixies. But, remarkably, the old crone's ever-ready help transforms the woman into a toad and the man into a frog. Animals of both land and water, toads and frogs are animals of transition, and by virtue of the many developmental forms that they pass through, serve as symbols of transformation. Indeed, one can think of them as advertisements for the possibility of transformation. But when a man and woman can only encounter each other as toad and frog, their meeting is limited to the realm of sexuality; individuality plays no role. Thus, after the water ebbs away, they lose track of each other again.

Now the man and the woman must follow their own paths of development, enduring many years of sheep-herding in grief and longing. This long period apart can be understood to represent the alienating rift that occurs after two people have loved each other merely as creatures of a species, so to speak, and allowed the longing for the human and the spiritual aspect of love to go unanswered. At this point, both partners retract into themselves and their own desiring, mourning—the tending of their own flocks. This is a work of concentration, of holding oneself together, which leads to outward as well as inward composure, and contrasts starkly with the boundless emotionality triggered by the nixie. In general, it appears to me that an emotional realm is now being cultivated that is quite the opposite of the world of the nixies. This realm is characterized by quiet longing, relatedness to oneself, and contemplation—which of course points to the principle that whoever propagates and lives out the "loud" emotions will have at some point to cultivate the "quiet" emotions. Examples of this principle are provided by many representatives of the so-called '60s generation, who, after preaching and living out the "loud" emotions in their early years, went on to become dedicated meditators later on.

Toward the end of the tale, springtime breaks forth, suggesting that the forces of life emerge again from of the earth, that eros lives again, thus that the husband and wife can be reunited. Although they don't yet recognize each other, a rapprochement takes place. And the full moon shines again. In general, the moon symbolizes the rhythm of all natural life. In this folktale it is clearly also to be seen in connection with the experience of the feminine. The full moon would then be the phase in which the positive aspect of the feminine comes to fruition. Now the man plays the flute—that is, he can reveal and express his feelings, the entire scale of his feelings for his wife. This touches her feelings as well, and she cries. Now they really recognize each other. Their relationship can be whole, for they are no longer swept away by life's instinctual undertow, but have made room for a dimension that is more refined, emotional, and intellectual.

If we consider the folktale once more as a meditation on anxiety and modes of handling anxiety, we notice that from the very beginning the narrative is burdened by the threat of the nixie—a relationship to the feminine, and therefore to sexuality, that confronts us with nature in the raw, engulfs us, fills us with longing, and pulls us away from reality. At first, anxiety can only be fended off by avoiding the danger; later, the development of facets of the personality not immediately related to the danger allow one to cope. As long as the nixie is only a fantasy, she poses no immediate threat. But in the moment that the relationship between man and woman becomes real, the threat of the nixie also becomes real. In the beginning only the man suffered the anxiety, but soon the woman, who was deprived of relationship altogether, had to bear it. By developing her nixie nature, she pulled the man away from his absorption with nixies. After many long years in solitude, longing and grief, the two at last find a relationship that can succeed. Gone is the fear of the nixie and the longing that dislodges the self.

Here, we have viewed the folktale as illustrating a relationship in which the woman, through her own inner development, helps the man out of his impasse. However, I do not mean to suggest that the man should be let off the hook. He, too, must develop himself (by "tending sheep") independently of the woman. Naturally, one could also interpret the wife's process in terms of the hunter's "anima-ego."

Getting Through Symbiosis

■ Introduction

In symbiosis, one person merges so completely with another person, group, country, or other entity, that any distinction between the two seems unreal. Those who are caught up in such a symbiotic relationship have the feeling that they are being cared for, protected, and relieved of the torture of eternal decision. Yet this is hardly a calm shelter, for to keep it intact requires continuous, anxious attention. Those who are in a symbiotic knot intensely fear that their "relationship" will break apart. "Relationship" is not really the right word for this attachment, for a real relationship requires two distinct individuals. On the contrary, someone involved in a symbiotic partnership does not dare exist as an individual. Fear of isolation and separation is the very stuff of which the symbiotic bond is composed.

"Symbiosis" is a term taken from biology that refers to a narrow, functional relationship between two organisms that is of mutual benefit to both. The symbiotically bound partner benefits from the knowledge that he is taken care of, protected in his helplessness, relieved of the pressure to make decisions and take risks. The "host" benefits from an enormous increase in self-importance, a narcissistic "boost" provided by the symbiotically attached partner. The essence of the relationship is not so much a matter of dependency as of not being separate, for the one who is symbiotically attached can just as well be the dominant one. Neither host nor partner can distinguish self from other, or establish which are his wishes and which are the other's, or differentiate his own ego from the other's.

What is the experience of symbiosis like? A twenty-five-year-old woman sums it up it quite succinctly when she describes her relationship to a girlfriend ten years older than she: "I no longer know whether it is I who think something or she. I buy the same clothes that she does. I know they don't really look good on me, but whenever I go shopping, I just can't get interested in anything else. I even changed my area of studies to coincide with hers, and now it really upsets me that I don't know whether or not I really am interested in the subject. When I am not with her, I feel completely lost. The slightest demand on me makes me feel like I am falling into a thousand pieces. When I am with her I feel confident again. She knows this, and so she can boss me around as much as she wants and I will put up with it. I can't stand being alone, and I can't stand being with her, not to mention the fact that she doesn't really want to be with me anyway."

This relationship is so symbiotic that its distressing aspects have completely eclipsed its positive ones. The young woman can no longer enjoy the feeling of being taken care of and the sense of omnipotence that come from merging because she has squarely faced the need for separation, which is why she sought a therapist. Nevertheless, she says that she feels confident when she is with her friend, and she feels she is falling apart when she is not with her. Thus we see how symbiosis helps to offset the threat of fragmentation. Notice, too, how much the young woman's ego flows into her friend's. Such dissolving of borders is frightening, but at least she can still observe and describe it.

A thirty-year-old man describes a very different kind of symbiotic relationship: "In the spring, when I lie down on the forest floor, I can feel all of nature pulsing through me. And when I am able to sense what is going on inside me, I feel surprising energies arise within. I am no longer myself; I am all of nature. I am the power of creation, everywhere and full of strength.

The sun shines through me. The smell of the earth and the forest floor blows through me. I could scream with delight and desire. My skin is too tight around my body; my spirit needs to stretch out and melt with my surroundings. It is a wonderful feeling. Nothing in my life is as intense as this. Sex is stale by comparison. Why can't it always be spring?"

This type of symbiosis empowers, though only for a moment. The young man in question entered therapy complaining of states of exhaustion that he "couldn't explain." In fact, his symbiosis with nature had imprisoned him. He could not escape it to return to normal human life, to be with other people. He only wanted to merge with nature, especially with the resurgence of spring, and when he could not, he would sink into apathy.

Finally, let us turn to the symbiotic dream of a twenty-three-year-old man, a student who changes his area of studies every semester. He experienced the dream as one of the most frightening and illuminating he had ever had.

> I am walking along University Street with some student friends. They are talking about a lecture, and I am trying very hard to understand what they are saying, but I can't. Max, a friend of mine, asks me why I am moving so slowly. I was not aware that I was moving more slowly than the others. Max stops and tells everyone, "Take a look at Freddy and his father: a real chip off the old block." Suddenly I feel my back caught on something— it has become fused with someone else's back. It's a warm feeling, but it certainly restricts my movement. I am terribly embarrassed that all of my friends can see me like this, but I cannot tear myself away; the connection is too strong. My friends huddle up to decide whom to ask to do the operation. They think it is absolutely necessary to separate us.

The dream speaks for itself. Fused with his father's back, his own back is strengthened, but at the cost of his flexibility. When his friends see this and decide that separation is necessary, the dreamer is greatly embarrassed. The dreamer himself is passive except for his slowly dawning awareness of the situation. Yet something within him—the "friends"—knows that a separation is called for.

The problem of symbiosis has been a classic issue in psychology. Jung wrote about *participation mystique*,[13] a concept adopted from the writings of the anthropologist Lévy-Bruhl.

> What he meant by it [participation mystique] is simply the indefinitely large remnant of non-differentiation between subject and object, which is still so great among primitives that it cannot fail to strike our European consciousness very forcibly. When there is no consciousness of the difference between subject and object, an unconscious identity prevails. . . . Then plants and animals behave like human beings. . . . Civilized man naturally thinks he is miles above these things. Instead of that, he is often identified with his parents throughout his life, or with his affects and prejudices, and shamelessly accuses others of the things he will not see in himself. (Jung, *CW* 13, par. 66).

Jung considered the dissolution of this mystical participation to be a great therapeutic achievement (ibid). He went on to say that the dissolution of *participation mystique*, of the identity of subject and object (or of symbiosis), comes about by means of *individuation*, a process which is seldom, if ever, finished. Individuation is understood here as "a process of *differentiation* (q.v.), having for its goal the development of the individual personality."[14]

Thus the way out of symbiosis, according to Jung, is found by "becoming conscious" of one's own identity, a process that is extended over one's entire lifetime. I prefer to speak of "symbiosis" rather than *participation mystique* or "identity of subject and object" because I think that it more clearly suggests a form of unconsciousness that takes place and causes difficulties within relationships. In short, I see symbiosis as a *type of relationship*.

"Symbiosis and individuation" also refers to research conducted by Margaret S. Mahler and her coworkers. Mahler often described a phase of "normal symbiosis," beginning approxi-

mately with the infant's second month of life, in which the
infant behaves as if it and its mother "were an omnipotent system—a dual unity within a common boundary."[15] The symbiotic phase is followed by the phase of separation, during which
a rapprochement between infant and mother also takes place. In
my view, the life-rhythm Mahler was describing does not apply
only to infants. Throughout the life span, phases in which the
individual seeks symbiosis are invariably followed by phases of
separation and individuation. Here, "separation" means emerging from a relationship characterized by fusion, and "individuation" indicates attitudes that have been acquired and that portray the individual characteristics that have been adopted.[16]
Within the phase of separation, a phase of rapprochement often
takes place, as if to reassure one that there is always a way back.
After this, there arises once again a need for symbiosis, this time
at another level. This need should be heeded, for an optimal
symbiosis is the prerequisite for separation and individuation—
in the life of the infant as well as throughout the life span.

Though optimal symbiosis is not my theme here, it is, in a
sense, a part of my theme. For "ways out of symbiosis" are easier to find if one's symbiotic needs have been met and reflected
upon. Beginning with the small child, it is obvious that optimal
symbiosis is a prerequisite for optimal separation and individuation. It would also be easy to misunderstand my title,
"getting through symbiosis," as implying that symbiosis must
be overcome at any price. This is of course not true. The very
reason I want to speak of "ways" out of symbiosis is that so
often one finds no "way." Sometimes one "sits around" in symbiosis. At other times, a phase of symbiosis simply lasts too
long, or is not allowed to evolve into a phase of separation. This
is the locus of many psychic difficulties we deal with in the
practice of psychotherapy: developmental inhibitions, identity
confusions, lack of creativity, depression, suicidal tendencies,
addiction. In extreme cases of regressive symbiosis, a person

may wish to return to the womb or simply to die—wishes that may be expressed in suicidal tendencies or psychosomatic problems.

Symbiosis and death have much in common. Our collective images of the next world are colored by our symbiotic needs. This is evident in the idea of paradise, for example, or such expressions as "entering into eternal bliss," "being absorbed into a greater wholeness," "resting in peace," or "reclining in the arms of God." Granting the uniqueness of each person or religion's conception, we must admit that there is a general tendency toward symbiosis in our fantasies about what comes after death, at which time our individual existences as we have known them on earth will cease. Some fear this, others long for it, and still others create fantasies in which the individual's uniqueness, indeed his or her isolation, is not completely abolished after all. Death and symbiosis are also related in that symbiosis always attempts to prevent changes in life. Fear of change and anxiety about having to move on and take one's leave—ultimately, fear of our mortality—prompt us to seek something as solid as bedrock. Often, we fall into the trap of seeking something much too solid—namely, symbiosis.

In every case of protracted symbiosis, one can ask, what development is being postponed here? Symbiosis may prevent us from facing up to life, which demands that we be reborn again and again, that we risk doing something new, that we never cease making new decisions in order to discover who we really are.

In the pages that follow, I will use folktales to describe particularly lethal aspects of symbiosis. I see the need for symbiosis, the danger of persisting too long in symbiosis, and the need to separate and individuate, as typical human issues, and I believe that folktales provide us with examples of ways to resolve them.

As I was searching for folktales that describe ways out of symbiosis, I was struck by how many tales depict "heroes" whose symbiotic needs hold them back. A Finnish folktale called "The Wife"[17] is one example.

THE WIFE

Once there was a man and a woman who lived together in peace and harmony and who were as fond of each other as it is possible to be. One day the man said to his wife, "When I die, you will surely find another man."

"And you will surely find another woman," the wife replied. "You will not stay single forever."

But neither the man nor the woman believed the other, and so they decided to make a pact that neither one would remarry should the other die.

As it happened, not long after that, the woman died. At first the man lived without a woman because he had no desire to marry again. But after some time had passed he thought, "Why should I go on mourning like this? I'm going to get married again." And indeed, he soon found himself another woman. Just as he was about to go and meet the bride, who was waiting for him by the church, he had an idea. "I shall visit my wife, bid her farewell, and ask the dead woman to forgive me," he thought. And so he went to her grave, knelt down, and said, "Forgive me! I am going to a wedding. I am marrying again."

At this, the grave opened up and his wife's spirit called out to him. "Come, come. Don't be afraid, come here," she said, beckoning him. "Don't you remember what we promised each other? That whoever survives wouldn't get married again?" Then she talked him into climbing into the grave with her. "Will you have some wine?" she asked him as they were both sitting in the coffin, offering him a glass. After drinking, he wanted to leave, but she begged him, saying, "Stay here a little while. Let us have a good talk!" Then she poured him a second glass of wine and he drank again. After this, he stood up to go again, but she insisted he stay and talk longer. "Do not go yet," she implored. And so the husband lingered on.

Meanwhile, the pastor was holding a prayer service at the church, for it was assumed that the man had fallen to his death. The bride-to-be, having waited and waited, finally returned to the home of her parents.

After the man's wife gave him a third glass of wine, she finally allowed him to leave the grave. "Go now!" she said. So the man left. When he arrived at the church, the pastor was no longer there. All the wedding preparations had been taken away, and he himself had grown as gray as an old hoopoe. He had been in the grave for thirty years.

The folktale describes a couple who live together in peace and harmony. They love each other so much that their only problem is that this paradise cannot last forever. The message is clear: things should be kept the way they are—so nice and free of aggression. Even death—the change of life— should not be allowed to interrupt.

To my mind, the initial situation in this tale depicts a typical symbiotic scenario: nothing should change, the aggression that could create distance (and thus change) seems to be perfectly eliminated, coming to expression only in their fantasies of death. And then the woman really does die. The symbiosis is broken by a higher power; the couple has no choice but to separate. The man grieves as befits a separation—especially a separation from what was apparently such a paradise—but after a while, he desires to be with another woman. This desire expresses the possibility that life can go on, if changes are accepted. The woman who has died could represent a real woman or an essential aspect of life that, for some reason, has faded out or died. In the tale, the husband portrays one way out of symbiosis: bereft of his object of symbiotic desire, he mourns, but after a certain amount of time has passed, he turns his attention to something new.

But when the time comes for him to marry this new woman—when the bond to the new phase in life really gets serious—he wants to say goodbye again to his old wife and beg her forgiveness. Naturally, he leaves his new wife "standing in the rain" at the threshold of the church. Now his symbiotic relationship returns as strong as ever to suck him back in. The fact that he must beg his old wife's forgiveness shows that he feels guilty toward her; he is not yet ready to do what he wants to do. His wife may be dead, but she is not inactive! Her grave opens up, she summons him in to remind him of their promise . . . and the old symbiosis is alive and well again. Despite the man's attempts to get out of the pit, she holds on to him until it is too late. The wine and conversation have sent him into relapse!

The pull exerted by the old situation is tremendously strong. It is common wisdom in myths and folktales that one should never accept food or drink from the dead unless one wants to go and live with them. The most famous example of this is Persephone, who is stolen from her mother, Demeter, by Hades, god of the underworld. After Demeter petitions them, the gods decide that Persephone can return as long as she did not eat anything while she was among the dead. But since she has eaten the seed of a pomegranate, she has to spend half of the year in the underworld.

In the Persephone myth, as in our folktale, the dead appear to seek the company of the living. Perhaps this portrays the fantastic pull of intense regression. Here, the bridegroom takes the wine offered to him. Wine is the traditional drink of the gods. In the cult of Dionysus, wine was thought to grant immortality and thus symbolized the power of the spirit to overcome all that is earthly. But of course, transcending earthliness can simply mean one ignores reality. By the same token, the man's conversation with his wife was intimate, confidential, and probably inspiring. Nonetheless, none of it got out of the grave. In this, it resembles the sort of extended dialogues some people hold with partners who died long ago, and are long gone as far as everyone else is concerned. Erich Fromm probably would have used his term "necrophiliac"[18] to describe such a love for the dead, which prevents the lover from attending to his living relationships. The problem is not the *relationship* to the dead, but the *nature* of that relationship. After all, though we do have to sort out our relationships with those who have died, we need to attend to other relationships as well. In general, it is wise to avoid any and all cults that would exclude us from life.

The married couple in this tale stay together even unto the grave ("till death do us not part"); by remaining faithful to his wife, the surviving partner is cut off from life. His relatives declare him dead.

One often finds a situation like this among partners and relatives of persons who have committed suicide. They draw out to great lengths their concern with the "unnecessary death" and are always looking for someone to blame (whom they naturally find). All they can do is think about their dear departed ones and all that they will miss about them. Lost to the world outside the grave, they are incapable of viewing their feelings of guilt from any other perspective than a literal one. For example, they cannot entertain the thought that perhaps guilt is necessary and unavoidable, serving to keep us bound to each other to some extent. The grave is often described as a kind of mother. We speak of someone returning to "the womb of death" or to "mother earth," as if to imply that a transformation takes place in the grave. But in our folktale, such a transformation cannot take place. The dead cannot "bury the dead" and what is over is not really done with. Instead, the husband gets comfortable with the past and has an intimate conversation with a significant bygone. He lacks aggression—not of a destructive sort, but an aggression would give him the verve to act decisively. For decades, he just sits there in the grave.

In other words, he never really lived his life. It's a complaint psychotherapists often hear, especially when there is a question of a symbiosis: "I never really lived my life; I always did what everyone else wanted me to do. . . ."

In this folktale, we have a man who lives as if he had already died. Instead of excusing himself and departing from his wife in a way that might border on abandonment, he destroys his entire life. We have all heard of people whose lives resemble this folktale—those who, out of a sense of "fidelity" to their deceased partner, refuse to remarry. Or those who are very deliberate about remarrying and yet insist that their new spouses do everything exactly as their old spouses did, and who keep the old portraits hanging on the wall . . .

But I think that this story can mirror other sorts of symbiotic clinging as well: to persons, groups, and objects, even when these are already "dead" or when the repressed aggression that could have broken the symbiotic attachment instead "killed" the object of symbiosis. Staying in symbiosis naturally offers protection—after all, who would have thought of looking for the man in his dead wife's grave. He was left in peace—buried alive.

"The Wife" illustrates several key aspects of the phenomenon of symbiosis:

• Prohibitions against change that are instituted in the symbiotic system extend even beyond the bounds of life. There is an overemphasis on what is shared, an avoidance of what divides.

• Aggression is split off or absent.

• Aggression that is split off (but may reappear in fantasies of murder) may cause the symbiosis to collapse.

• There is evidence of the beginning of a process of working through the loss: mourning, accepting the new situation, caring for what is new.

• The old symbiotic tendencies reawaken, one is reattached to the "dead" partner, the symbiotic system is reestablished, though not with the same intensity as before. Old and new are mixed together, resulting in a withdrawal from all social contacts and present concerns for an extended period. Rather than removing oneself from the symbiosis, one destroys oneself.

Of course, in actual life symbiotic tendencies may not have such dramatic effects as they do in this folktale. However, I still think the tale accurately illustrates a typical and unmistakable dynamic of all such relationships. As I have said, the alternation of symbiosis and individuation comprises a rhythm in everyone's life. No one is simply "symbiotic." In all of our lives, certain spheres are characterized by symbiotic tendencies; these tendencies are perhaps the strongest in those areas in which we feel most helpless. Or perhaps it is better to say that we are most helpless in those areas in which we find ourselves in symbiotic attachments. Then again, these areas present us with some of our greatest opportunities for individuation as well.

We seem to dwell somewhere between our need for symbiosis and our striving for separation and individuation. Let us not forget, however, that individuation out of symbiosis is never easy, and involves not only feelings of separation but of abandonment. Such painful feelings may dog us from earliest childhood on, never vanishing completely no matter how proud we are of our steps toward independence.

Such feelings of separation and abandonment can push us back into symbiosis or, as Fromm strongly advocated, they can be suspended through love. The movement out of symbiosis toward greater individuation, which happens many times in life, becomes ever more integral with the addition of the dimension of relatedness and love. "In contrast to symbiotic union, mature love is a unity under the condition that one's integrity and independence are maintained, and thus also one's own identity."[19]

"Mature" love is not something one can seek and find, but something that the personality must grow into and develop toward. Yet it is clear that the love Fromm called for—this relatedness between two persons who are striving to develop their own individual identities—only becomes possible when one can recognize and suffer through one's symbiotic tendencies on the one hand and one's urge for individuation on the other. Mahler and her coworkers repeatedly emphasize that the discovery of identity depends on both optimal symbiosis and optimal separation and individuation. Blanck and Blanck are of the view that problems encountered in the first phase can, to a degree, be corrected in later phases.[20]

To Fromm's concept of love I would add the idea of relatedness, for I do not think the two are identical. I believe that we can learn relatedness, and even demand it of each other. Love, by contrast, cannot be demanded, for it involves an element beyond the grasp of the will. Relatedness means we resist seeing others as mirrors of ourselves and attempt to understand them as unique personalities, with their own fears and wishes, interests and energies, which we react to and contend with. Relatedness should increase the chances that two people can arrive at a shared point of view that each can affirm without forfeiting his or her identity.

But beyond relatedness, it is love, which cannot be manufactured, that takes hold of us, leads us beyond ourselves, and ends our previous isolation. This sensation can also be experienced in sexuality, which elegantly expresses the rhythms of merging and returning to autonomous individuality.

To conclude these introductory remarks, I would like to consider briefly various situations in which symbiotic tendencies are likely to arise. Or perhaps it would be more to the point to talk about situations where such tendencies are not apt to arise. As we have said, symbiotic tendencies are "normal" for infants. Here it is usually the mother who symbiotically merges with the small child. This original symbiosis can broaden out in the course of life to include one's family, persons whose views are similar to one's own, one's homeland, race, religious community, and so on. This is not to say all such relationships are necessarily symbiotic; they may be symbiotic, inspiring a fierce inner loyalty and causing one to lose one's critical capacities and become hostile toward those who are different.

We can symbiotically merge with anything that promises us a degree of protection and prompts us to surrender our independence. Thus it is only natural that we would tend to merge symbiotically with our partners. For years it was maintained that women were too symbiotically attached to their mothers to achieve the independence required to form new kinds of partnership. Here the symbiotic bond to the personal mother is seen as the defining characteristic of women in general.

Therapists would do well to acknowledge that symbiotic merging takes place in many therapeutic relationships—indeed, that it needs to take place if the client is to find the shelter and safety that is the prerequisite for separation and individuation. But we must also be cautious not to feed this symbiotic merging with our own reluctance to let go of our clients as they move toward greater autonomy, or to give up a relationship that boosts our narcissistic sense of our own importance. The client's bid for symbiosis calls for a decision on the part of the analyst; depending on the situation, the latter may even choose to fight against the analysand's attempt to draw him or her in. After all, "ways out of symbiosis" include ways out of the healthy symbiosis in therapy as well as ways out of dangerous therapeutic situations.

Mystics through the ages have discovered a most unique form of symbiosis—one which involves merging with the Divine. However, this longing for fusion with God—which they invariably describe as a longing for death—has not kept most mystics from doing their work among people—for the sake of God, naturally. Teresa of Avila was exemplary in this regard. She described a mystical experience as follows:

> While I was reciting it [the hymn, *Veni, Creator*], there came to me a transport so sudden that it almost carried me away: I could make no mistake about this, so clear was it. This was the first time that the Lord had granted me the favour of any kind of rapture. I heard these words: "I will have thee converse now, not with men, but with angels." This simply amazed me, for my

soul was greatly moved and the words were spoken to me in the depths of the spirit. For this reason they made me afraid. . .[21]

This experience, which could be characterized as one of merging with a celestial being, did not stop Teresa from working hard on earth. Indeed, I don't think we could accuse Teresa of being merely symbiotic and not individuated. Perhaps true mystics live out their symbioses in a way that does not interfere with their process of individuation. Perhaps they have found the optimal way to be symbiotic.

All in all, then, we must say that symbiosis is neither good nor bad in itself. In certain situations, it is extremely important for individuals to be cared for symbiotically, so that new energies can be brought to life in them. But just as often, symbiosis can cause an individual to withdraw into total estrangement from life. It may mark the onset of sterility and, by preoccupying us with death and the past, obstruct the onward flow of life. I have attempted to describe the natural rhythm between symbiosis and individuation, and suggested that our efforts be directed at maintaining this rhythm in such a way that each new phase is characterized by increasing differentiation. Through such a process, we acquire a greater sense of personal identity and begin to experience our separateness as an opportunity for a true alliance based on relatedness and love.

Still, all along life's path, we encounter situations in which we must escape or abandon symbiotic bonds. In the following pages, we will examine such situations: ways out of symbiosis that are also ways to separation.

■ Journey to the Underworld through the Hellish Whirlpool of Fafá

CONTENDING WITH THE
DEVOURING PRIMAL GROUND

Lauango and his wife, whose name was also Lauango, had three sons—Faalataitafua, Faalataitauana, Faalataitimea—and a daughter, Sina. One day the two older brothers went fishing. Faalataitimea, the youngest, stayed home, and was entrusted to care for his sister. After he made dinner for her, she lay down and fell asleep. Sometime later, she awoke with the realization that someone had caused her a terrible disgrace. Just in that moment, she saw her brother going out the door. She rose up feeling miserable and went out onto the beach. There she sat down and waited for her two older brothers to return from their fishing expedition.

Quite some time passed before the boat approached the shore. "I think I see Sina sitting there on the beach," said Sina's eldest brother to his partners. "I wonder why she is just sitting there in the harsh sunlight. We had better row quickly." When the brothers got closer to land they raised their voices, "Yes, it really is Sina!" When it was time for the fish to be unloaded onto the shore, Sina asked her oldest brother to excuse her from helping out; she wanted to go out to sea again with Faalataitauana in order to catch some one-eyed tunafish.

And that is what happened. The two got into the boat and rowed away. When they were far out at sea, Faalataitauana began to wonder what Sina really had in mind. "Don't you want to fish? There's plenty of fish here," he said. "Oh no," Sina said. "We better keep rowing. It's too sunny here. That's probably why you didn't catch any good fish before."

After they had rowed even farther out to sea, her brother again said, "Sina, there are such beautiful fish here! Let us fish!"

And once again, she answered, "No, no, keep on rowing!" And on and on they went until they reached the entrance to the underworld, where rushing waters plunge in a mighty racket into the hellish whirlpool of Fafá.

"Before I die," cried Faalataitauana to Sina over the din, "I want to know why we are both about to plunge to our deaths."

"Yes," sighed Sina, "And I want to tell you. It is because that stupid little boy Faalataitimea disgraced me forever!"

"What did he do to you?" asked her brother. "Tell me and I will kill the miserable joker."

"Come on now," answered Sina, "we are almost in the whirlpool. See that rock at the edge? See the tree on it? You can still save yourself by jumping out of the boat and grabbing onto the tree. As for me, I'm going to go down into the thunderous depths."

Faalataitauana did as his sister had instructed him and grabbed onto the tree. Meanwhile, Sina was seized by the foaming, gurgling whirpool and swallowed up. Then Faalataitauana climbed onto the shore and lay down to sleep on the beach under the creepers, which belonged to Sisialefafá, a very honorable lady. Once the boy was asleep, birds began singing all around.

When she heard this wondrous sound, Sisialefafá went to the shore to find out what was happening. Seeing the beautiful, slumbering youth made her practically lose her senses for joy. "I'm going to scare him to find out if he is a human being or an evil ghost," she thought, and cried out, "Hey there!"

The noise startled Faalataitauana. "What's going on here?" he said as he got up. "Why did you startle me like that?" "Ah" said Sisialefafá, "so you are a chief. Tell me, why are you sleeping here under the creepers like a beggar? I have a nice house with sleeping mats, nice headrests, and good mosquito netting."

"I saved myself from the ocean," the youth answered, "and I had to rest on the shore."

"Come with me!" said Sisialefafá, and she led him to her house. After she had given him fine clothes, mats to sleep on, and a comfortable headrest, Faalataitauana went to sleep some more. Meanwhile, she went outdoors and heated up some cooking stones on the fireplace. Then she took two taro fish and two hens. She cleaned one taro but left the scales on the other, she plucked one hen and put the other in the oven with its feathers still on. When everything was ready, she put the unplucked hen and the uncleaned taro on a banana leaf and served it to Faalataitauana.

"Dear woman," he said when he woke up, "would you be so kind as to take this food away? No one eats this sort of thing where I come from. We scrape the scales off our taro and pluck our hens."

"Oh, please forgive me," said Sisialefafá, "my people have made a mistake." And so she took away the first meal and came back carrying the nicely prepared hen and the taro that had been cleaned.

Now they both sat down to eat. "So, what is your name?" Faalataitauana asked when they were finished.

"I am Sisialefafá!" she replied. Then he told her his name, too. Sisialefafá's hair hung down to the ground, for it had never been cut. And Faalataitauana said, "Hold your hair in a bunch, I want to cut it off!" She did as he said, and after it was cut they went bathing in the ocean. Then they went back to the house, where they lived happily together for some time. Sisialefafá bore him one child and then another.

One day Faalataitauana began to pine for his sister. "I want to know where I can find my sister Sina," he told his wife. "She was swallowed by the Fafá-whirlpool."

But Sisialefafá suspected her husband had different motives. In her jealousy, she said, "Oh, so you want to have an affair with Ilalegagana?"—referring to an elegant lady who lived near the hellish whirlpool and was secretly in love with Faalataitauana. Sisialefafá knew that Ilalegagana had powerful love magic, which she had found no antidote for. And so, seeing that her husband was determined to find his sister, she made him an apron of good titi grass and gave him a splendid necklace of red pandanus fruits. "Put on this apron and necklace," she said. "They will protect you and my love against Ilalegagana." Then she instructed him about her rival's love magic, telling him about the little shells which Ilalegagana had named after Faalataitauana and his brothers.

And so Faalataitauana went to the whirlpool. As he approached, he saw a group of young girls standing around. "My, look at the beautiful chief coming over here," they whispered to each other. Could Faalataitauana, the one whom our lady is always thinking of, be as beautiful as he?" But Faalataitauana paid no attention to their gossip. All he cared about was getting his sister back again. But to get to her he had to go through Ilalegagana. First he had to destroy her love magic. So he went in her house, grabbed one of the enchanted shells that she had hidden behind her back, and broke it into pieces. Ilalegagana cried, and uttered a protest:

"My little shell is broken and gone,
the little shell that I called Faalataitafua;
I carried it hidden behind my back,
He is Lauango's oldest son."

Then he grabbed the second shell, which Ilalegagana kept between her eyes, and broke it, too. Again she cried and protested:

"My little shell is broken and gone,
the little shell that I called Faalataitauana,
Like Lauango's second son, who is always on my mind,

I always kept it where I could see it."

Now Faalataitauana grabbed a shell that was lying on Ilale-gagna's lap and broke it. And yet a third time Ilalegagana cried and uttered her protest:

"My little shell is broken and gone;
the little shell that I called Faalataitimea,
Like Lauango's third son, the idle little boy,
I had it hidden here on my lap."

Faalataitauana threw the pieces down at her feet. "How could you give my name to such a lousy little shell?" he demanded. Though Ilalegagana herself was silent, the people in her house would not endure this insult to their lady, and so put up a terrible ruckus. But Ilalegagana wanted reconciliation, and so told her servants, "Go and get all the beautiful things that you have made ready for my love; the pig, taro, yams, hens, coconuts, sugar cane, and the slaves."

When all these had been delivered to Faalataitauana, he turned to the servants and said, "My people, why are you bringing me all these things? What should I do with them? It would be better if you divided them among yourselves!" So half of the gifts were given to those who were present and Faalataitauana took the other half.

The love magic of Ilalegagana proved to be strong; Faala-taitauana forgot his Sisialefafá and ate with Ilalegagana, who became his wife. When they were finished eating, the young man said, "Now I would like to have a look at Vaiola, your fountain of the water of life!"

"But who told you about the fountain?" asked Ilalegagana. "Who has instructed you in this matter?" Suspecting that it was Sisialefafá, she ordered her people to go and capture her and have her baked in the oven. They soon returned with Sisialefafá and her two children. Sisialefafá knew what was about to happen and lamented:

"Calm and soothe yourselves, please,
soothe and calm yourselves, please.
Let me tell you how it all came to pass,
how Faalataitauana appeared
slumbering on the strand,
under the creepers,
and about the song of birds,
And how I went to the shore to see
if he were a human or a ghost.
How I stood there,
while, swimming, he saved himself
from the foaming sea.
About the children I bore him.
Tauana, come forth to me,
To you I give my children,
even while I am burned."

Even Ilalegagana was moved by this lament. "Let Sisialefafá and the children live!" she proclaimed. They should return in comfort to their home." Still, she kept Faalataitauana for herself.

One day Ilalegagana said to her husband: "You wanted to see Sina again? Go now and find the place where the old, blind Matamolali lives. Break off a coconut leaf and touch her eyes with it." So Faalataitauana went out and did as he was instructed. When he touched the old woman's eyes, Matamolali called out, "Oh, who is touching my eyes?" And in that moment, her sight returned. In exchange, Faalataitauana asked for a favor: to be led to the entrance of Vaiola, the spring of life. Matamolali agreed and went with him to open the way.

After a while, they came to a broad river, where they saw figures with arched backs floating on the waves. Next came people with rotten limbs, swollen arms, and bloated legs, then people who had lost their eyesight; and finally the bodies of healthy men in fine shape.

But Faalataitauana did not see Sina anywhere, so Matamolali went to the one who had guided Sina into the underworld and asked what Sina had looked like. "When she sank with the boat," her guide replied, "she was wearing nothing but a necklace of red pandanus fruits." As he said this, a parade of virgins was carried over the waters, the last of which was Sina. "Dear," Matamolali called out to the girl, "please bring me your necklace!"

"Here," whispered Sina, holding the necklace out.

"Bring it to me!" said the old woman.

"Oh, forgive me," replied the girl, turning to go. "I have to go now; the others are waiting for me."

"Bring it here right this minute," the old woman demanded, "or else I will lose my patience and shut down the springs. And then there will be no more swimming and romping about!"

And so Sina came toward her with the necklace, and Matamolali grabbed her by the hand and yanked her out of the waters of death. Then the old woman took Sina to Vaiola, the waters of life, which flowed close by. She smacked the girl and dunked her under the water, then smacked and dunked her once again. Then she asked, "What is that over there?"

And Sina answered, "That is the west."

The old woman smacked and dunked her again. "What is that over there?" she asked.

"That is the east!" Sina said.

And again the old woman smacked and dunked the girl. "What is that over there?" she asked again.

"That is the south!" answered the girl.

One last time the old woman hit and dunked. "What is that over there?"

"That is the north!" Sina said. And thus she was brought back to life.

Afterward, Sina went with the old woman back to her home, where Faalataitauana was waiting in a locked-up part of the house where no one could find him. Matamolali gave the girl a

in society. Our interpretation will perforce be restricted to a description of the tale's most important features, and will necessarily omit many of the tale's uniquely South Seas features. After all, if folktales do describe typical human problems—and solutions to them—then we should be able to find in a South Seas folktale lessons that are valid for humanity in general.

The folktale begins with Lauango, who has a wife by the same name. With this lack of differentiation between father and mother Lauango, the theme of symbiosis is suggested. However, rather than focusing on the parental situation, our folktale focuses on the youngest brother's violation of his sister in her sleep—an act that "causes her terrible disgrace." This may indeed be a consequence of the parents' symbiotic relationship: if the parents share a single name, why shouldn't brother also sleep with sister? Then everything stays "all in the family," and the original symbiosis is not destroyed. Or so it might seem. Actually, the youngest brother's act destroys the symbiosis. His sister, who had been sleeping, suddenly wakes up. Suddenly, she is aware of what is happening: feeling violated, she exposes herself to the scorching sunlight. With Sina's outraged realization, the scene changes, the plot begins. Her waking up launches her, and the situation, onto a path of development.

The initial situation in this folktale might be compared with a family in which the members, knitted together in symbiotic bonds, sleep a great deal—that is to say, much occurs on an unconscious level. One day, a brother decides to satisfy his sexual urges by means of his sister. Perhaps he chooses her because he knows her and is afraid of other women. But brother-sister incest also often reflects a brother's feeling of dominion over his sister. He wants to have exclusive rights over her; he wants her to serve only him. In our tale, the brother exaggerates his symbiotic tendencies, thus initiating a developmental path. He has broken a taboo whose very existence tells us how strong is the temptation to go back to the sister rather than forward to another woman.

Of course, in this interaction, Sina herself is ignored and has nothing to say. Thus it is surprising that immediately after it, she becomes very active. Now she is determined to go fishing out at sea with one of her brothers. But this is only a pretense; what she really wants is to sail to the underworld, "where the rushing waters plunge in a mighty racket into the hellish whirlpool of Fafá." Sina wants to die. It seems to her that nothing short of this radical solution will work. She cannot live with her disgrace, and feels that death alone will absolve her of it. But as we know, death can be understood as a grand form of symbiosis. When defeated in obtaining their original symbiotic goal, people with symbiotic tendencies may choose death as the "next best" thing, rather than suffer through an unsatisfactory situation one step at a time, and thus develop out of it.

Sina's brother asks why the girl is steering them toward imminent death—a question that saves his life. For as she careens into the thunderous depths, she instructs him to hold onto a tree on a rocky cliff beside the whirlpool. Being seized and pulled down into the swirling depths is a vivid image of symbiotic merging, in contrast to which we have the image of the tree by which her brother is saved. The tree is a symbol whose many aspects all point in the direction of life. Especially in tropical or torrid regions, the tree indicates the presence of water, which is so crucial for survival. The tree is also a phallic symbol, which could indicate that redemption is forthcoming through greater efforts toward autonomy. The advice to hold on to the tree comes from his sister, and thus, rather than disappearing wordlessly into the whirlpool, she insures that something in life will survive. But it is the brother who inserts the "why"—this one small question that aborts an unconscious symbiotic process that is careening toward death. And indeed at this point in the tale, he becomes the locus of development.

Before proceeding with our interpretation, we must ask ourselves if the brother is really the brother—i.e., someone who, though his sister has initiated his development, proceeds here-after largely independent of her? Or, should we view him as the masculine side of Sina, the folktale's heroine? Alternatively, we could view Sina as the feminine side of Faalataitauana. These sorts of questions always crop up regarding the interpretation of brother-sister folktales. The question is particularly pointed in the case of this folktale, and particularly difficult to answer, because the tale itself is so symbiotic, with so many actions merging into others. I am going to propose that we consider Faalataitauana the hero of the tale, but that we acknowledge the possibility of other viewpoints from time to time.

Once again on the theme of symbiosis, it is important to note that at this juncture a separation takes place between Faala-taitauana and his sister. Faalataitauana is abandoned; he loses something—seemingly forever. Applying this situation to life, we are reminded of a symbiotic relationship that has come apart, leaving one partner to find himself suddenly alone. Having lived so much of his life in the other, it is as though part of him has died. He may have saved his neck one more time, but too much has been lost of the life he once knew. Thus do people with a basically depressive psychic structure react to separations. Having tied themselves symbiotically to their partners, perhaps even making them the "bread of their lives," they suffer an intense sense of abandonment once the other leaves.

Faalataitauana now lies down in the creeping vines to sleep. Here again, we have the motif of sleeping—i.e., symbiosis. As the youth lies down in Sisialefafá's creeping tendrils, one knows it will only be a matter of moments before he is embraced by Sisialefafá herself. Creeping vines suggest the entwining, entangling, and engulfing aspect of the mother archetype, the naturely aspect that extends itself outward. At the same time, there are the birds, signaling Faalataitauana's pres-ence. Thus, a celestial, spiritual domain has been activated as well as the earthy, vegetable domain of the mother archetype. Then there is the matter of Sisialefafá's name, which suggests we are still in the region of the hellish whirlpool, which is called Fafá. Thus we might say that Faalataitauana remains in the realm where he lost Sina, even if he has forgotten her for the time being.

Sisialefafá now ensnares him decisively—and, it must be said, without much resistance from him. He is spoiled with pleasures and seems to enjoy a veritable paradise. And yet he does not let himself become inextricably entangled in Sisiale-fafá's snares: he rejects the food she has not cleaned properly, i.e., that which is uncultivated and uncouth. And he cuts Sisiale-fafá's long hair, which has never been cut before—an episode that reminds us of the story of Samson in the Hebrew Bible.[23] From that famous story we know that there is power in long, uncut hair—for women, it is an erotic power, which adds so much to their power of attraction over men. Hair that has never been cut also suggests a woman who lives in primal conditions without contact with human civilization, or one who lives in a still-unconscious region of the psyche. Once this long hair has been cut, both partners dip into the ocean. Bathing, washing, and being immersed in life-giving waters form an important cluster of motifs in this folktale. Bathing suggests a washing away of this paradise-dwelling Adam to clear the way for a new phase of growth.

Having understood Sisialefafá as a female figure that belongs to the mother archetype—part mother, part lover—we must ask what it means that Faalataitauana stays with her. On a very concrete level, we could say that we are dealing with someone who has lost his place of symbiotic belonging, and who flees directly into another symbiotic, seemingly maternal relationship so that he can cease worrying about what he is missing and enjoy instead what is served to him on a silver platter. But we must not forget that Faalataitauana's collapse into symbiosis is not complete; he does maintain some degree of autonomy. This woman is powerful, but she does not have complete mastery over him.

Such a symbiotic condition can as easily exist in one's imagination as in a real relationship: we see it reflected in fantasies of paradise, of being courted and taken care of. Notice, however, that Faalataitauana does not merely fantasize passively; rather, he takes an active stance. Instead, we might call this a compensating fantasy, one that arises in reponse to a major loss and that, in my view, stems from the mother archetype.

I am reminded of a thirty-year-old man who sought therapy because of his depressed moods. These had begun on the occasion of his twin sister's marriage. Before then, both siblings had lived together in the same house. (The brother had been quite spoiled.) On the surface, the young man was glad his sister had married, for he had always resented the way she hovered over him and treated him like a child. But once she was really gone, his life seemed intolerably boring, and he toyed with the idea of ending it. Soon after he began therapy, a mother figure appeared in his dreams whom he named Mother Holle. In a sequence of dream images, he was fed and bathed by her. The unconscious satisfied his needs; indeed, he did entertain such fantasies. Yet I felt that it was important for him not simply to be lulled to sleep by the more or less seductive women in his dreams; he needed to remain alert and conscious every step of the way.

One day Faalataitauana begins to pine for his sister. Having waded through his grief and taken up his life where it had been broken off (the proof that his life has gone on lies in his children, whether real or imaginary), he now realizes that something essential is missing. He has not progressed after all, nor will he until he gets this essential thing back into his life.

In spite of her jealousy, Sisialefafá tells Faalataitauana what to do to get to Sina. He must go through Ilalegagana, a woman who lives near the whirlpool and thus will bring him closer to the problem of being engulfed. Secretly, this woman is also in love with Faalataitauana. Though this detail may reflect a bit of local color finding its way into the story, I feel it expresses something crucial about how one finds one's way out of symbiosis. That is to say, more is required than the power of the man's own will; fate herself must seek him out, and he must accept his fate.

To get to his sister, Faalataitauana must go through Ilalegagana, and Sisialefafá can instruct him in this woman's magic ways. She carries three shells, each of which embodies one of the brothers. By gathering within himself the potential of all three brothers, Faalataitauana becomes capable of untying the knot of family symbiosis with which the tale began. Why shells, we might ask? Shells carry and protect something precious, but they also enclose; thus the image suggests entrapment and deprivation of contact with the outside world. The shells in this story are further described as "little," and indeed they do seem miniature in the presence of this great mother who carries her sons on her body like little toys, guarding them fiercely. The grand and overwhelming power of the mother archetype becomes grotesquely clear. But, thanks in part to his training with Sisialefafá, Faalataitauana has become so autonomous that he knows he must undo this miniaturization of himself and his brothers: he grabs the shells and smashes them. "How could you give my name to such a lousy little shell?" he asks, and so reveals that he has become conscious of his worth and will no longer settle for this brittle husk of a life.

Taking a stand and insisting on his own worth gets results: Ilalegagana bequeaths him a great many articles. He does not keep these exclusively for himself, but gives away half to the others present. Goods and energies—unrealized life possibilities—that had been in Ilalegagana's keeping are now at Faalataitauana's disposal. The fact that he chooses to share them seems to me a statement about the broadening of his personality. And yet he still succumbs to Ilalegagana's wiles.

What might this mean in terms of a relationship? As I see it, the movement toward Ilalegagana represents a step toward greater consciousness. Faalataitauana realizes something—perhaps that he is in a relationship with a woman who does not value him any more than she would a shell she wears for adornment around her neck. More than infantilized, he knows he could be crushed at any moment. Or perhaps—moving to the realm of fantasy—he suddenly realizes that he is as vulnerable as a fragile shell worn by a goddess named Fate, and that he must put up a fight against her, resist her, refuse to accept her. If he did not, he would resemble those depressed persons we often see, who experience themselves as little more than pawns of fate, who have lost all sense of their ability to initiate change, and who have consigned themselves fully to the charge of others. Here Faalataitauana has become strong enough to reject this demeaning situation; he stands his ground with dignity and deals decisively. It also makes sense to me that, after his victory, he would be seduced by Ilalegagana: an act of autonomy is always followed by an act of symbiosis.

Ilalegagana does not recover so easily from her defeat, but looks around for someone to blame. I think this scene gives us a clearer picture of what Ilalegagana is all about: a kind of Mother Holle of the South Seas, she is clearly a manifestation of the mother archetype in which the dimension of the lover, though still present, has faded into the background. His meeting with her suggests that, psychodynamically, Faalataitauana is coming closer to the core of the problem. In this regard, Sisialefafá represents a level of development to which he must not return, but which, nonetheless, he must allow to live.

Finally, Ilalegagana sends Faalataitauana to visit the blind Matamolali, who knows the location of the entrance to the fountain of life. With a coconut leaf—a remedy from this world —he is able to restore the old woman's sight. This remarkable image of the blind mother who sits beside the fountain of life brings to mind the expression "blind fate." Exploring this thought further, we are reminded of Ilalegagana, who had put her shells in places where she could not see them: between her eyes, on her back, in her lap. Evidently, seeing—a mode of conscious perception—has not been very important until now; indeed, in symbiotic situations, seeing is never the most essential thing. On the contrary, it sets the seer apart from what is seen, which is exactly what a person in symbiosis fears most. If Faalataitauana can restore Matamolali's sight, some potential that had remained in darkness can be realized, something denied and closed off from the world can now become visibly active in the search for the fountain of life.

It is my conviction that in working with a complex, it is not only the person who is doing the work that changes (as in our story Faalataitauana changes in relation to his various women), but something in the unconscious also changes (Matamolali opens the gates of the fountain of life). An ancient symbol for the life-giving womb of life, the fountain is the place where the earth's riches and overflowing bounty spill out freely. Here the water necessary for life is brought from the depths to the surface, where human beings can put it to use. Our tale is explicit about this source, from which life once again surges forth. The souls of the dead pass by this stream of living water. As in so many other tales, the dead here are symbolic representations of various possibilities that never found their way into people's lives. The task at hand is to win back the possibility represented by Sina, who must not be allowed to disappear again. Matamolali pulls her out of the water of death and conducts her to the water of life. Thus we see that Sina was not irrevocably lost after all; she can be brought back, but only by Matamolali.

Next comes the strange ritual of submersion, beating, and testing for knowledge of the four directions. What could this mean? Submersion in the water of life is not difficult to comprehend. Depressed and longing to die, Sina had been immersed in the waters of death; now these must be rinsed away. Nor did she come willingly out of the deathly waters. Beating in folktales often signifies an attempt to dislodge something, to get someone to give up or let go of something. Such logic is probably also behind the idea that beating is a necessary part of child-rearing. One also finds, with beating, the intent of drumming something into a person's head—of getting them to wake up and notice something. Here, it is the four directions that are impressed upon the girl. The old woman begins with the west, the direction associated with the realm of the dead, and does not end until she has drummed all the directions into Sina's consciousness. Such symbolic orientation is necessary for survival in the land of the living, and Sina can only return to life when all of the four directions have been beat into her head.

People like Sina, who emerge from a deep regression, are usually quite confused, and need someone or something to provide them some basic orientation. In our tale, Sina is reminded of her brother by means of his belongings. So in life can personal possessions that were significant before the onset of a severe regression—whether a psychosis, a major depression, or a state of withdrawal—act like signposts, indicating the way back into daily life.

And now brother and sister find each other again; what was dead comes back to life, it is time for a return to the homeland. Until now the entire story has taken place in the hellish whirlpool, which reminds us of ancient Greek entrances to the underworld. The abysmal waters of the river Styx were described as a waterfall that crashed over a rock wall some 200 meters high. Oaths were made by these waters, which acted as a witness unto death. Indeed, the waters of the Styx were considered so lethal that they could only be carried in a hoof or a horn.

In our tale, the waters of life and death are not far apart. This suggests that the problem of symbiosis is worked out in a psychic sphere in which success and failure are very closely knit. Although the female figures that come forth are all extremely helpful, the whirlpool of death is never far away. Faalataitauana's development probably took such a favorable course because he constantly refused to let himself be sucked into the whirlpool, and remained committed to the world of the living. Therapy often requires a similar strategy. Though we may be well aware of the great danger of a given situation (such as a patient's potential to fall into a deadly symbiotic trap), we do not rivet our focus on the trap. Instead, we concern ourselves steadily with those areas of life that promise growth and provide the patient with psychic nourishment.

Now brother and sister leave the realm in which life and death are so close, in which Faalataitauana discovered the fountain of life, and return to their homeland. At this point, all the gains that have been made, all of the inner development that has taken place, must stand the test of actuality. Otherwise another symbiosis would form—such as an all-too-familial fixation with inner progress—that could easily turn into infinite regress.

The sun rises (another symbol for the resumption of life), brother and sister return home, and Faalataitimea, the evil brother, is found dead, punished by the gods. Thus we are told that the brother who wished to remain within an incestuous symbiosis no longer exists, and life can take up its onward course again. Here is where a new folktale should begin, with Sina as the active figure, becoming involved with the son of a foreign chief.

I have interpreted the tale in terms of a brother-sister relationship. Out of the initial situation, which resulted in an act of incest between brother and sister, the necessity arose for contending with the factors underlying this symbiosis. Though

Sina conducts her brother to the place of confrontation, she soon disappears, leaving it to the masculine to wage this confrontation.

The action described in the tale has all the earmarks of a confrontation with an engulfing mother complex. The devouring aspect is portrayed in the image of Sina being swallowed by the whirlpool, suggestive of a deep regression. And yet a confrontation with the mother archetype need not always result in one's being swallowed up completely; one can also let oneself be only partially engulfed. Thus we see that as the protagonist gains broader autonomy, he learns to avoid the disempowering influences that the women in the tale represent, so that he may be led by them to the fountain of life. In this way Sina, who represents an essential psychic content formerly lost to him, is brought back to life. The part of him that died is revived. The experience of passing through death and rebirth leaves a deep mark and conveys a sense of hope, of possibilities constantly awaiting us. The realization of what has been gained here will take place in the homeland, the place where the problem began.

This folktale can be understood on a number of different levels. It illustrates the dynamics of a couple in a symbiotic relationship, each of whom has a serious mother problem. For example, in a relationship between two depressed persons, the mother problem would be in desperate need of attention. In facing the problem, there are no guarantees. One possible outcome would be a fall into an even deeper regression, a yet deeper symbiosis such as is portrayed here with Sina's disappearance in the hellish whirlpool. Another possible outcome would be the gradual acquisition of autonomy, gained through perseverence. Clients working through such problems may need to use their analysts as if the latter were all-giving mothers with whom it is safe to risk their first steps toward autonomy.

It seems to me that the two poles represented by the brother and sister in our tale do not refer to sex roles so much as two responses that are ever-present and fluctuating in any emotional situation. At one moment one falls into the trap, and at another one preserves that which nourishes.

But the tale can also be viewed from the perspective of Sina, starting off with the same symbiotic situation. In this case, we would see the story as a description of the conditions under which someone could wander off the deep end of his or her commitment to life. It would illustrate certain suicidal tendencies, literal or psychic—the latter being common among symbiotically oriented people, who give up quickly and do not give themselves a second chance. Viewed in terms of Sina, the story addresses the same underlying mother problem, which still needs to be uncovered. But now we would say the problem is tackled with the help of the heroine's masculine side, which until then had not been autonomous enough.

A question naturally remains to be answered: Can symbiosis be characterized as an issue belonging to the mother complexes? And if so, can a course of therapy that provides an experience of the positive, nurturing aspect of the mother archetype without falling into its traps, show the way out of symbiosis? The question certainly deserves further exploration. We will return to it at the conclusion of this book.

■ Redhair Greeneyes

A WAY OUT OF ATTACHMENT TO THE FATHER

Once there was a merchant who told his son, "My boy, if I die, don't, under any circumstances, hire 'Redhair Greeneyes!'" Time passed, and the light went out of both his eyes.

"My son," the father instructed, "go and get yourself a servant so you can keep the business afloat. Go now and make hay."

On his way into the city, the son saw two men throwing a corpse out of their house and then beating it. "He owed us money which he never paid. Now he's dead and gone. We are beating him because of the money he owed us."

"If I pay his debts, will you leave him alone?" inquired the son.

"So be it," answered the men.

And so the lad paid the debt that was owed, buried the body, and went on into the city. After wandering around until evening without finding anyone to take into his service, he went home.

The next day a man stopped him on the street and asked, "Sir, do you need a servant?"

"Yes, I do," the lad replied. However, seeing that it was Redhair Greeneyes, he had to turn him away. "I am sorry," he said. "You are not the one I am looking for."

After he had wandered around a bit more, he returned home and told his father what had happened. "Tomorrow," the father instructed his son, "go into the city and make sure you find a servant."

When the lad went into the city the next day, Redhair Greeneyes approached him again. And so he decided to take

Redhair Greeneyes home with him. When he went to see his father that night, he explained, "I wandered around everywhere, but I couldn't find anyone, so I took Redhair Greeneyes."

"Well," said his father, "if you have already brought him home, I guess that is the way it will be."

On the following day, the lad got up, prepared his things, packed his wares, and set out with his new servant. Whenever they paused during their travels, the servant made tea for his master, tended to the animals, made his master's bed, and drew his sword to watch guard over the merchandise.

One day they came to a crossroad where an old man was standing. The youth greeted the man and then addressed him, asking, "Old man, we are on our way to Damascus. Of the three roads going from here, which one leads to Damascus?"

"Your honor," replied the old man, "all three roads lead to Damascus. If you take this road, it will take you six months and you will arrive safe and sound. If you take the other road, it will take you four months, and you may or may not arrive. By that other road there, it would take you only two months, but as yet no one who has traveled it has returned."

"Servant," the lord said, "we will take the six-month road!"

But Redhair Greeneyes said, "No; we will take the two-month road!"

"Ah," the merchant said to himself. "My father told me not to take Redhair Greeneyes as my servant, but I didn't listen to him."

Failing to persuade his servant, who had his mind made up, the lad agreed to take the two-month road. After traveling for two days, they took a rest at nightfall. Redhair Greeneyes unloaded the pack animals, prepared a meal for his master, made his bed, and drew his sword to guard the merchandise.

Partway into the night, he heard their dog barking and saw that a dragon had come to the tent. "Hey dachshund," the dragon called out, "why doesn't someone kill you and take out your brain? If a blind man were to put it on his eyes, he would see again."

"Hey dragon," the dachshund responded in kind, "why doesn't someone kill you and grind up your skull-bone? If a man who had been decrepit for seven years were to put it on his body, he would be healed!"

Raising his sword, Redhair Greeneyes went after the dragon and, with one stroke, let the head fly. Then he brought the head back and put it in a box, which he closed and placed along with the merchandise.

The dragon was the reason why no one who took this road had ever returned.

The next morning, Redhair Greeneyes loaded up the goods and called his master, who mounted his horse and rode in front while Redhair Greeneyes took up the rear. They rode together until they reached Damascus, where they rented a room and set out their goods for sale.

One day a town crier came around. "The king has rheumatism," he announced, "and has promised to whomever can heal him whatever his heart desires."

Redhair Greeneyes said: "Lord, you will say that you can heal him."

"Lad," said master to servant, "what do I know about medicine? I have no experience in such matters."

"I'm telling you," ordered Redhair Greeneyes, "either you do it or I'll hack you in two with my sword."

"Good, I will heal him," said the merchant, because he feared Redhair Greeneyes.

The town crier then brought the message to the king. "On such and such a street," he told him, "there is a merchant who says he can heal you."

"Lad," the king commanded the crier, "bring him here." On their way to visit the king, Redhair Greeneyes instructed his master, the merchant. "Take this dragon's head and grind it up in a mortar. Mash it up well and spread it all over a roll of unbleached cotton. Undress the king and wrap his body in the cloth. Have him lie there for twenty-four hours, and then remove it again."

The merchant did as his servant had instructed. When the time was up, he removed the wrapping and the king stood up as healthy as on the day he was born.

So the king said to the merchant. "Come here and tell me what it is that your heart desires."

"Your majesty," the young man answered, "there is nothing that I need, for I am already rich in possessions. However, if it is your desire to please me, then give me the hand of your daughter."

"My dear young man," responded the king, "in the name of God—if only you had asked for anything in the world other than this. In return for the good deed that you have done me, how can I give you my daughter? Dear fellow, I have already married her three times, and each time the marriage lasted   only one night. The next morning her husband was always found dead."

"Well, they may be dead," the merchant replied, "but I still want your daughter." And so the king gave him his daughter.

That evening a bed was made ready for the bridal pair. Redhair Greeneyes drew his sword and kept watch over the young couple after they fell asleep. Soon he saw the girl's braids beginning to shake. Thrashing back and forth, they turned into snakes. When they wrapped themselves around the bridegroom's neck as if to strangle him, Redhair Greeneyes struck their heads off with his sword. That night the king could not sleep. "Oh God," he moaned, "this young man did me such a good deed, giving me my life back. Please let him still be alive come morning!"

The next morning at the break of dawn the king sent some of his people to see what had happened. "Go, and bring me good news," he bid them.

"We bring you good news," said the men upon their return. "Your son-in-law is safe and sound."

At this, the king ordered a grand celebration that lasted for seven days and seven nights. He assembled twice as many goods as the young man had and gave them to him, saying, "You are most welcome; now go your way."

With their wares in tow, Redhair Greeneyes followed his lord and his lord's new spouse. Whether they went a long way or a short way, they traveled until they reached the outskirts of their city. Redhair Greeneyes said, "My lord, you know that all of this belongs to me."

"Yes," the lord acknowledged.

"Let us divide it up fairly," said Redhair Greeneyes.

"As you will," responded his lord.

After dividing the entire lot into two parts Redhair Greeneyes asked, "Lord, is there anything else?"

"No, what should there be left?"

"What about our dachshund?"

"Let him be either yours or mine," answered the merchant.

"No," said Redhair Greeneyes. "We will divide him up."

"For God's sake." said the merchant, "You can have him. What shall I do with him?"

"It is a matter of fairness. I will split him in two."

Thus, Redhair Greeneyes drew his sword and split the dachshund in two, from head to foot. "There," he said, "Take whichever half you like, my lord."

"You know what is best," said the merchant. "It is between you and your God."

"Rightly so," said Redhair Greeneyes. "This is for you and this is for me. Is there anything else, my lord?"

"There is nothing else," he said.

"Oh yes there is; there is your wife."

Now his lord was deeply dismayed. "For God's sake," he said, "Let her be yours or let her be mine."

"No," said Redhair Greeneyes. "I demand that justice prevail. I will cut her into two pieces, one for you and one for me."

"For God's sake," cried the merchant. "How can you cut her into two pieces?"

Redhair Greeneyes drew his sword and stood over the girl, who shrieked in fright. As she screamed, two snakes slithered out of her nostrils.

The servant attacked and killed both of them. "My lord," he said, "these two snakes were the work of your spouse. They could have caused you harm. Thanks to me, you wife is safe now. Take the dachshund's head home with you. Grind it up well and put it on your father's eyes and they will be healed. As for me, I am going to die now. For I am that man whose body you saved from my debtors. I prayed that God might allow me another three months to live in order to pay you back for your kindness. Now I must die. But if you would, please bury me here and let God watch over your ways."

Then Redhair Greeneyes died on the spot. After he was placed in the ground, the merchant gathered together his goods and went home with his wife. He healed his father's eyes and told him about how Redhair Greeneyes had done nothing but good for him.

As they went on with their lives, their wishes were fulfilled. May your wishes be fulfilled as well!

––––––––––

This folktale, from Kurdistan,[24] depicts yet another type of symbiosis, which requires its own unique way out. The tale begins with a merchant and his son. Apparently there are no women in this household, or if there are, they are not important enough to be mentioned.

In making plans for his son for after his own death, the merchant denies him his autonomy. This strikes me as a symbiotic situation. But the father does not die here, he "only" goes blind

—which could signify his incapacity to see Redhair Greeneyes's importance for his son and for his own future as well. Becoming blind means losing foresight and perspective. Images of the past remain, but no new ones are created. Naturally, the motif of the blind seer is also suggested—the old man who, having lost his outward sight, sees great truths within. Perhaps this father has to look within in order to see that he is trying to protect his son by keeping him from his own experiences. Or perhaps the most important thing about the merchant's blindness is that it causes the son to take on a servant, around whom the story revolves.

The son is extremely passive and never does anything unless his father asks him to. We find such a figure in many folktales: i.e., a son who is bored at home and asks his father for money in order to see the world or find a woman. Such a figure remains a son to the end, an extension of his father, who remains the master.

In many such tales, as in this one, the issue is the son's separation from his father and acquisition of autonomy. From a more sociohistorical perspective, it concerns the upcoming generation's renewal and redefinition of paternal and masculine energies. If we see the father and the king as representatives of the masculine principle, we see just how badly such rejuvenation is needed: the father is blind and the king is rheumatic!

On his way into the city, we are told, the son encounters two men beating the corpse of a man who died before he could repay his debtors. The dead man still owes something to the living; he has failed to live up to something important in his life. The corpse represents an aspect of life that was excluded from the symbiotic situation. As the target of beating, it brings us face to face with the aggressive parts of the personality. This display of aggression is hardly one that enhances the quality of life. The men who beat the corpse are preoccupied with something that has long been dead. This imagery makes me think of grown-up sons and daughters who insist their parents still owe them something. Of course, as much could be said of most of us—which is perhaps not such a bad thing. After all, if our parents had given us everything, we would hardly have the chance to discover anything new. Still, a good many sons and daughters spend their entire lives beating on this "corpse" in the belief that it would come alive if they just beat long and hard enough. This is, of course, a futile exercise which drains one's energy. Ultimately, it proves nothing except that the children expect to be given what they cannot earn or accomplish for themselves. This is, indeed, a demonstration of symbiotic demands.

The son in the folktale does exactly what is required in such a situation: He pays off the debt attached to the corpse, redeeming whatever the person failed to deliver in life, and buries the corpse, throwing dirt over it. He placates the problem and pays for his father's omissions. We all pay for our parents' omissions, and our children will pay for our omissions as well. Here, the young man lays the problem to rest. Perhaps we can see one more aspect of the beating in light of this: sometimes we bury our problems before we have recognized and suffered through our feelings about them. Then again, we can also be too slow to bury our problems. Here the son seems to come along and bury at the right time, putting an end to the matter so that life can continue. Funerary rituals can be seen as a means for transforming the dead: by "reminding" them that they need to return to the earth from which they came, they are aided in rejoining the cycle of fertility. Thus we might say that for his first step toward autonomy, the son takes the initiative to ransom something that had been excluded from the symbiotic system, some bit of unfinished business. He pays attention to an issue that was neglected under the old regime.

In his search for a servant, the son finds only one candidate, Redhair Greeneyes, whose offer of help he at first rejects. But on his third day of looking, having no other alternatives, he decides to take the man his father had explicitly warned against. And yet when informed of the choice, his father is surprisingly quick to agree. From the end of the tale, we know that Redhair Greeneyes is none other than the corpse that the son ransomed. And having ransomed him, he cannot resist him. The son has set free an entire realm that the corpse possessed away from life, and now this whole realm must be integrated into life.

What realm might the figure of the servant portray? A hint might lie in the name assigned to him. Red and green are opposite colors, suggesting not only a tendency to clash but the ability to arrive at a certain balance. Red hair and green eyes: if the figure were a woman, such a combination of attributes would suggest she was a witch. Red hair indicates a fiery temper and passions, a capacity for reacting with sharp affect and deep emotion. There is danger present, but also warmth. In the symbolism of folktales, red hair—like red beards—are attributes of the devil. But the fact that we are dealing here with demonization does not tell us much; we need to ask what is being demonized. Could it be passion itself that has been associated with the devil, an impetuous life-energy that does not balk at aggressive outbreaks? Clearly this "red" element has been lacking in the son's upbringing—otherwise his father never would have sent him away.

In contrast to his red hair, the servant has green eyes. Capable of gazing deeply, green eyes are mysterious and difficult to read. Such eyes seem to hide some secret, indeed, we may even worry about them becoming poisonous. His green eyes, together with his dynamic and aggressive appearance, give the servant a shady quality that makes him seem like a man with underground connections. As with the father, the eyes here are significant. However, Redhair Greeneyes very definitely "sees" —and perhaps he also disapproves.

All forms of passion, all relations to the shadier regions of life, have been lacking until now. As long as the son was nice and did not stray from his father, life remained as lifeless as a corpse. But now the very embodiment of the "other side" has been employed as a traveling companion and given a leading role in all deliberations along the way. Now he cannot be ignored. Though his companion is hardly a father figure, the son is still a long way from making decisions; Redhair Greeneyes retains the authority. Though the son would have chosen the safest way to Damascus (in keeping with the initial symbiosis), Redhair Greeneyes chooses the riskiest: the two-month road, from which no one has ever yet returned. He chooses the way of possible death, a way that may well lead to ruin.

Redhair Greeneyes can be understood as a psychic element in the son that has a great deal of eagerness and capacity, having just been ransomed from the dead. The ego can do little to stop this inner dynamism, which is something that we are aware of experiencing when others tell us, "Slow down; you are ahead of yourself." We ourselves can only try to keep up; our determination concerns even us.

Redhair Greeneyes chooses the path of risk and confrontation. Knowing that no one has ever returned from this path, he can expect to encounter a decisive danger there. By taking the six-month road, he would been guaranteed of a safe arrival in Damascus, but he would not have been transformed. And the purpose of a path is to provide an *experience*, a significant event, rather than to preserve safety and the status quo.

It does not take long for an event to take place; in the night, a dragon appears. Not satisfied with a meal, the dragon seeks a quarrel with the dachshund that has accompanied the merchant and his servant. Dragon and dachshund each advise the other that they are capable of killing him, but also that their respective brains can be put to good use—the brain of the dachshund for curing blindness, that of the dragon to reverse a general decline in health. As we know, Redhair Greeneyes kills the dragon, removes his head, and then awakens his master.

The dragon belongs to the world of monsters more than the world of human beings. One imagines such a creature as a winged snake, a monstrum that belongs as much to the earthly as the celestial realm. However, its wings do not free it completely from earthly concerns; usually dragons are appointed the task of guarding a treasure. Generally they watch over this treasure so effectively that a great deal of courage, strength, or cunning is required to get to it; in taking on the dragon there is a good chance that one will be consumed.

One can view the dragon as something profoundly unconscious in the human psyche, something that wants to swallow up everything that is newly conceived, to destroy every step taken forward in consciousness, to send one regressing back to some archaic level. But conquering the dragon liberates a treasure.

The dragon in our tale is a talking dragon, which indicates that it is much closer to the world of human beings than many. This is a dragon that can be dealt with. Indeed, Redhair Greeneyes kills it with very little difficulty. And we would probably all agree that it needs to be killed, that the danger of severe regression it represents must be eliminated. But the dachshund also needs to be killed. The dachshund is a hunting dog trained to track badgers. If we regard the dachshund as one of the son's behavioral possibilities, we might say it is his doggedly faithful side, the part that follows the orders of his master—or his father—without thinking.

Both of these behavioral possibilities arise out of symbiosis and the repression of Redhair Greeneyes, and both can be worked through. For now the issue that was denied has been accepted into consciousness, indeed, it now plays a leading role in its development. The dialogue between the dragon and the dachshund is similar to an intrapsychic dialogue between complexes. Once a complex has been recognized, it often begins a spontaneous process of recovery, which can be witnessed in dreams and fantasies.

To illustrate this spontaneous process, I would like to use the following dream. The dreamer was a young man whose self-esteem was extremely unstable as the result of his father's authoritarian style.

> A flashy guy about my age, looking cool and tough on his big Honda, meets a shy, skinny young fellow, and looks him up and down without saying anything. The skinny guy is afraid and holds on tight to his girlfriend. Suddenly the flashy guy gets off of his motorcycle and motions to the shy one that he can take a turn on the bike. The bashful guy asks his girlfriend if she wants to go with the motorcycle rider. She nods. The skinny guy sends her off with the flashy one.

Without going into the dreamer's associations here, we can note that the dream clearly illustrates two of the dreamer's behavioral options—flashy or shy. During the course of the dream, these extremes come closer together and indeed communicate with each other. The dreamer did not recognize either of them as a dream-ego; he simply saw them as young men about the same age as himself. Thus he showed that he was not yet conscious of these as potential modes of his own behavior. Split off from his ego and sense of self, they are patterns of behavior dictated by his complexes, reactions that come alive when he trips up or loses his balance. The rapprochement that takes place in the dream—and thus in the unconscious—represents the complex's spontaneous process of recovery.

Returning to the complexes in our story, we are told that neither the dragon nor the dachshund should be completely eliminated. Their brains—i.e., their essence—should be preserved, processed, and utilized medicinally, following the adage that the correct dosage of poison can be curative.

Suffering from rheumatism, the king of Damascus must be quite stiff. According to a theory put forth in psychosomatics, inhibited aggression may be one factor in the genesis of rheumatism.[25] The king, the father of the land, is cursed with illness. In general, all the forces of the masculine seem to be weak and under attack—so it's little wonder that Redhair Greeneyes is nothing but a vapid corpse. Now he who goes with Redhair Greeneyes sets out to cure the king. Of course, he does not embark on this undertaking of his own accord, but only in response to Redhair Greeneyes's threats.

How is the king cured? The head of the dragon must be ground up and applied to an unbleached cotton cloth. Then king must undress himself, then wrap the cloth around his naked body and remain in it for no less than twenty-four hours. In this passage, we find a description of the "renewal of the king," an ancient rite of death and rebirth. In such rites, the old "nature" is taken off and the new nature put on (Ephesians 4:22-24). We know that in ancient times, unbleached cotton was used for wrapping corpses—hence the suggestion of ritual death. What's more, the pulverized dragon's head should effect a real transformation. If the king's head comes into contact with the essence of the dragon head, the king will absorb a measured dose of dragon nature—a process that should have a curative rather than a poisonous effect. In all, the scene describes beautifully a patient working-through of the negative, destructive potential the dragon represents. Such a working-through requires a perspective that allows the positive side of the complex—here, nondestructive aggression—to penetrate the king's pores as well.

I find that we become destructive whenever we deny the aggression associated with a highly directed pursuit and avoid the intense confrontations that may result. If we deny this aggression, we are often left with an "all or nothing" solution, which, more often than not, is destructive. In this tale, the king is not directly destructive; his aggression is repressed and delegated to his daughter. He must therefore be reeducated about his own destructiveness so that his healthy aggression can be mobilized to restore his flexibility and free him of his "rheumatism."

The king is transformed and reborn, and so is the son, who is now called a young man. And when it comes to his reward, the young man asks for the king's daughter. This time, it is not Red-hair Greeneyes who makes the decision, but the young man himself. The king's flexibility is restored, the masculine recovers, and the young man becomes more autonomous.

But his choice of reward appears not to have been such a happy one: as it happens, the king's daughter has already been married three times and the morning after each wedding the husband was discovered dead (cf. the story of Tobias in the "Book of Tobit" from the intertestamental *Apocrypha*). As we have seen, this is the result of the king's repressed aggression, which had been delegated to his daughter. But the three dead bridegrooms don't make much of an impression on the young merchant, who by now is squarely on the path of risk.

And so, when the young man and his new wife go to bed, the girl's braids begin to quiver and turn into snakes. But before the snakes can strangle the sleeping bridegroom, Redhair Greeneyes strikes off their heads. What danger do these snakes represent? We noted that there were no women present at the beginning of the tale who might have helped break the symbiotic bond of father and son. Now a female figure enters who represents a danger to men, and we see that "repressed" feminine energy and repressed aggression are two sides of a coin.

What does it mean that the woman's braids turn into snakes? In the form of a braid, hair that otherwise might be erotic is tied up and ordered, and thus becomes snaky. Erotic demands become sexual demands which endanger the young man's survival. *They are highly dangerous in more than one regard.* The passion that was split off and embodied in Redhair Greeneyes surely included sexuality and sensuality. Looking once again at the initial situation, it occurs to us that, had it been permitted, such sexual passion might have guided the young man away from his father's side and ended the symbiosis long ago. Now, sexuality can no longer be repressed. A transformation must take place, and transformations are always hazardous.

Already, the son is changed in that he is no longer simply the son of a father, but also the husband of a woman. But now he must consummate the sexual act in order for the full transformation to occur. This is one danger. Another is that sexuality always involves a moment of self-abandon, a willing fall into an orgiastic realm. Often experienced as a kind of death, self-abandonment triggers intense anxiety. I know of clients with unstable egos who do not wish to have an orgasm because they are afraid of losing themselves. But once Redhair Greeneyes has killed the snakes, the hero of our tale no longer has to face this danger. Having gained a certain amount of experience with emotions and aggression, he can actively face the potential violence of the sexual situation as well. It could be destructive for him if he lost his independence and became completely subjected to the woman, which often happens when eros and sexuality are involved.

The next day, this marriage of masculine and feminine is celebrated with joy—joy that the woman is no longer so overwhelming, joy that all men who become involved with her need not be condemned to death. As it turns out, the solution was not for woman to become weaker but for man to become stronger. This broadening of possibilities for all concerned is indeed a cause for celebration.

Then the young man is sent on his way, along with his wife and twice the amount of goods that he started out with—an indication of how much his personality has expanded. But before the travelers reach their destination, Redhair Greeneyes wants to divide everything up. At first he claims it all belongs to him, which is in a sense correct, since the merchant's good fortune only began after he salvaged the corpse and took this new servant into his service. Then Redhair Greeneyes asks that everything be divided up and differentiated. Up until this point, the merchant's relationship with Redhair Greeneyes has been very symbiotic, and indeed, it has been quite similar to the merchant's relationship with his father. Now the symbiosis that has been transferred must be dissolved. This is accomplished through Redhair Greeneyes's last task, namely, through establishing what belongs to him and what belongs to the merchant. A similar task confronts anyone attempting to sort out whether a power they experience within themselves is something under the control of their ego, or something that transcends their consciousness. If it is anything like Redhair Greeneyes, such a power probably will never allow itself to be integrated completely.

Making distinctions is no easy task, and the merchant clearly resists dividing up the dachshund. It seems he doesn't want to know what his half of the dachshund represents in terms of the unfinished business. To divide and to share, one must be able to stand back from the object in question and look at it realistically. Here, the merchant has a chance to become more conscious, but first he must become aware of his dachshundlike behavior, for it must be sacrificed.

As it turns out, the woman is to be divided, too—or at least subjected to a kind of exorcism that drives out the evil still present in her. It is noteworthy that the snakes exit through her nostrils. In addition to breathing and the exchange of outer and inner, the nose is associated with a capacity for sensing the future, as in the expression "she has a good nose" for things. The snakes remind us once again of the bride's destructive potential. Once again it is Redhair Greeneyes, the embodiment of passion's positive aspect, who fights with passion's destructive aspect. Having killed the dragon and the snakes in the braids, he now attacks the serpents in the nostrils.

If a passionate, dynamic element that had been repressed is suddenly "ransomed" and allowed to act freely, we must not expect its effects to be unambiguously good and constructive but must reckon with a certain destructive potential as well. Often, however, the positive aspect will create the conditions in which the destructive aspect can be successfully dealt with.

At this point in the story, Redhair Greeneyes dies—which tells us that the aspect of life that he represents has been integrated, at least partially. To a certain extent, the merchant is now capable of living like Redhair Greeneyes.

Now the young man returns home and restores his father's sight. Over the course of his journey, he has achieved something not only for himself, but for all fathers, including the king. When wholeness is restored, one need no longer seek its illusory substitute in a symbiotic relationship.

One might note that the end of the tale finds the son once again living with his father, which prompts one to ask whether the symbiotic attachment has truly come to an end. But let us not forget that this is an oriental folktale, which derives from a culture in which it was customary for several generations to live in close proximity. However, comparable tales make it very clear that the son has become the carrier of the family tradition. And besides, a son with such an intensely positive father complex would be unlikely to lose it altogether; most likely he would continue to live it out in relation to other father figures. In any case, it is safe to say that the symbiosis at the end of the tale is nowhere near as enveloping as the symbiosis at the beginning.

Let us attempt to schematize the way out of symbiosis portrayed in this tale. At the beginning, the son is a "victim" of a symbiotic relationship with his father. This arrangement is disturbed when the father goes blind, thus forcing the son to become more independent.

The son takes his first step out of symbiosis when he ransoms the corpse. Seeing what was excluded from the system, he pays for it and puts it to rest himself. The crucial fact here is that he acts on his own judgment, even if only out of common decency. This gesture then leads to an act of disobedience when he hires the very helper that his father warned him against. At this point the tale begins to resemble a number of "brother" folktales in which two heroes strengthen each other. In "Redhair Greeneyes," the figure who appears to accompany the son gives him exact information about what to do at each step along the way. In obeying him, the son renews his symbiotic tie to a father figure. In a therapeutic situation, this could describe someone with symbiotic tendencies who suddenly dreams of a powerful, authoritative figure who shows him that the way out of his problems is to develop capacities that had been stifled in the symbiosis. This dream figure may appear as a brother or a sister, and since the symbiotically bound person is prone to obey, may have quite a lot to say about how the transactions of daily life should be handled.

A thirty-six-year-old analysand provides us with a good example of such a directive inner figure. This man was closely bound to his father, was even employed by him, and dared to conduct no business without discussing it in detail with him. When the young man was away or on vacation, he called his father on the phone frequently, sometimes even several times a day. He came for counseling (after talking it over with his father) because, although he had met a number of women, he had not been able to build a lasting relationship. In his dreams, there appeared several times the figure of a mountain guide who was modeled after someone he had met once. The guide became very important to him as he showed the man how to scale steep cliffs and spurred him on to other risky adventures. The mountain guide embodied values such as adventurousness, courage, faithfulness, and simplicity. As the dream figure waxed in importance in the young man's life, the father waned; at the same time, the son's sense of his own significance and initiative also grew. Of course, in the end, it was necessary also to "separate" from the mountain guide.

Although identifying with this dream figure temporarily boosted the analysand's self-esteem, he was forced to admit that he himself wasn't anything like a mountain guide. His own unique personality was composed of quite different behavioral possibilities and values. The mountain guide seduced him to peaks of valor beyond his true sense of himself, to acts of courage that were necessary in the phase of separation, but that would not have been very meaningful for the remainder of his life journey.

Often it is the analyst himself who takes on this role of companion figure. In such cases, the phase of separation (or, as in the tale, "dividing") becomes quite tangible. When the companion is projected onto the analyst, an "inner companion" is usually constellated as well. This phenomenon appears throughout human history in the form of guardian angels and other beneficent spirits. Thus, we can see that the inner companion includes a dimension that transcends both analyst and client. Analysts cannot play the role of the celestial guide for long, nor can clients expect to realize its exalted qualities during their lifetimes. The companion, like Redhair Greeneyes, is a figure of destiny; part of him can and must be integrated into life, but other parts will and must be left out.

In the course of the story, the companion restores what was diseased: here, the realm of aggression and impulse. New realms come into the picture as well, such as the relationship with a woman. Our tale aptly expresses both the life-enhancing and life-threatening aspects of what had been repressed. Capable of swallowing life up, the dragon can also be used as a remedy; the woman who was a cause for joy is also the bearer of deadly snakes. In the course of encountering these figures, the son's autonomy grows.

As a next step, the symbiosis with the companion also has to be broken, and this requires exhaustive deliberations concerning what belongs to whom. When the initial symbiosis is finally dissolved, life can continue on its course.

One could also see the tale as a description of a complex that, split off from awareness at first, is finally returned to consciousness. Symbiosis is a system that serves to defend against the complex. Let us not forget that conflicts portrayed in folktales may also reflect the times out of which they arise. However, since the Redhair Greeneyes-type of tale has been circulating for a very long time,[26] we may conclude that it represents a universal human problem.

■ The Daughter of the Lemon Tree

A WAY OUT OF OVERPROTECTION

Once upon a time, there lived a king and queen who were known far and wide and who had only one child, a handsome prince. When the prince was sixteen, he would walk to school with the son of the vizier. They loved each other as if they were brothers. One day the two youths were playing on the roof of a small house in which there lived a poor old woman who was cooking lentils. Accidentally, the prince threw a stone that fell down the chimney and landed in the old woman's clay pot, breaking it to pieces. The poor old woman ran out of her house, and seeing the prince on her roof, called out to him, "As much as I wanted to eat the lentils I just cooked, so much should you want the daughter of the lemon tree!" This frightened the boys away, and since it was already getting dark, each returned to his own house.

From that day on, the prince began to fret. He could neither eat nor sleep, and all night long he thrashed about in his bed, moaning and groaning. The king grew very worried and called first the court physician, and then other physicians, and later still others. But no one was able to determine what had brought on the prince's condition. Finally the son of the vizier, who knew the cause of the disease, told the king everything. When they heard the reason, the king and queen promised their child that they would get him the daughter of the lemon tree, and the prince began to recover the state of health that he had formerly enjoyed.

The first thing he did after getting up from his sick bed was to pay the old lady a visit. First he begged her forgiveness for doing the damage, then he pleaded with her to tell him where he could find the daughter of the lemon tree. At first she would not tell him, but when he promised her that he would give her a great reward, she said to him, "Listen carefully, my child: The daughter of the lemon tree is far away. And if you are going to find her, you will have to get yourself three pairs of shoes made out of iron, each of which will last only one year. You will have to be on your journey for three years, alone, with no beast or coach to carry you. You must always travel eastward, toward the rising sun. You must keep on without stopping for rest. You will use the first pair of shoes during the first year, the second pair during the second year, and the third pair during the third year. By the time the third year is over, the third pair of shoes will be full of holes, and you will come to a marvelously beautiful palace. There you will find the daughter of the lemon tree." Filled with joy, the prince thanked the old lady, gave her the reward, and hurried home to tell his parents.

When they heard of his plans, the king and queen began to fuss and brood. They were afraid that they would lose their son if he set out on such an adventure and so tried everything to keep him from going. But his mind was made up. When he refused to give up his wish to go, they finally granted their permission, fearing that if they did not, he might get sick again. And so, while the prince set out on his trip in good spirits, the king and queen dressed in black and wondered if they would ever see their son again.

The prince put on the first pair of shoes and crossed over mountains, through valleys, over flatlands and rivers, through woods, up steep inclines and down steep slopes, without ever stopping. Only at night did he lay down to sleep, and early in the morning he was already on his way again. After a year of this, his first pair of shoes was full of holes, so he put on the second pair and kept on walking. After the second year, he put on his third pair of shoes and continued on his long trek. When he still hadn't found a castle after the third year, he began to lose faith. Dejected, he sat down on a rock and thought for a long time about whether to turn back or keep on going. As he was thinking, he watched the huge globe of the sun disappear beyond the horizon, when suddenly he saw something in front of him off to the right. It was a beautiful castle, which sparkled, shone, and glowed, as if it were made of jewels.

The prince's spirits picked up immediately, and he hurried off to the castle. When he got there and entered the gate, he saw a strange sight. There in front of him were a lot of people doing all kinds of handiwork, but no one ever glanced at the person next to him or said a word. Though he drew near and hailed them, no one looked up at him, responded to his greetings, or paid him any attention whatsoever. Greatly dismayed, the prince again lost faith. But in the midst of his hurt he suddenly heard a voice calling to him, "Good child, come here, up the stairs!" Gathering himself together, he ran up the stairs as quickly as possible, and there he saw a young man his age. After they greeted each other, the young man asked the prince why he had come to the castle. "To find the daughter of the lemon tree, that's why," answered the prince. And the young man replied, "You shall have the daughter of the lemon tree. But you cannot see her tonight. Eat well, sleep, and tomorrow we will meet again and discuss the matter."

Then the prince asked about the people he had seen doing handiwork below, and why it was that they could not answer his greeting. "They are making other people's fortunes," said the youth. "But I am *your* fate. So eat now, sleep, and you shall have your heart's desire."

And so the prince lay down in the soft bed provided for him, but no sleep came. Never in his life had a night seemed so long. When morning finally came, he got up, washed himself, said his prayers, and went into the dining room, where he found a table covered with all kinds of delicious things that had been put there by invisible hands. But he had no desire to eat; all he wanted was to see the daughter of the lemon tree!

Just then, the youth who was his fate appeared. "Eat that you may be strong, for the journey you must make is a hard one," said the youth, whose name was Fortune. When the prince had finished eating, Fortune clapped his hands, and suddenly there appeared a winged horse that had a human voice. Now Fortune told the prince that if he sat on the horse and did everything the horse told him to do, he would get what he wanted.

The youth then swung himself up onto the horse, gripped its flanks with his legs and held the reins tightly with his hands. Then the horse took off, and as fast as lightning—one, two, three—it came to a garden with a large tree that had three lemons on it and was guarded by forty giants.

As they were coming down to land, the winged horse told the prince about the giants. "The giants guard the lemon tree so that no one can steal its lemons. Inside each lemon is the daughter of the lemon tree. Hold on tight! I am going down now. When I fly past the tree, grab one of the lemons and don't let go of it, whatever you do."

The prince did just as the horse instructed. When he came up close to the tree, he grabbed one of its fruits. The tree began screaming and the giants went running about in a blind frenzy, unable to see who was disturbing the tree!

Lemon in hand, the prince suddenly found himself back inside the palace, where he was greeted by the young man who was the spirit of his fortune. No sooner had he dismounted than the winged horse sped away. Then Fortune said, "Slice the lemon open with care." When the prince cut tenderly into the lemon, a beautiful girl with golden hair leapt forth! The prince embraced and kissed her.

Fortune then told the prince, "If you were to return to your homeland on your own power it would take three years. So I will have the winged horse take you there instead." Then the winged horse appeared and the prince and the daughter of the lemon tree, whose name was Goldenhair, instantly found themselves back in the prince's homeland. All who saw them flying overhead were smitten with marvel and disbelief. When the couple dismounted at the castle, they were greeted by a splendid reception. The king and queen threw their arms around their son and the golden-haired daughter of the lemon tree. Their engagement was solemnly confirmed, the old king willed his son the throne, and the marriage was celebrated.

But the good and virtuous prince was not appointed by Luck to enjoy his beloved Goldenhair for long. Not long after the wedding day, a neighboring king declared war on the kingdom. Before the prince left for battle, he built a tall and sturdy tower in which he placed his Goldenhair, appointing a number of men to stand guard around it, and entrusting a maid to serve her faithfully.

While the prince was at war, the girl with the golden hair spent her days in sorrow. One day as she was sitting by the window, a deep slumber came over her. The wicked maid noticed this and seized the opportunity to push her out the window. Then the maid quickly ran down the stairs and disposed of the girl's body in a deep and muddy pond where no one would find her. But first she cut off the girl's golden hair and affixed it to her own scalp so she could be as beautiful as her mistress. Then, putting on her mistresses' golden clothes, she pretended that she was the princess, and no one noticed the foul deed that she had committed.

When the prince returned from the war, he went straight to the tower to visit the golden-haired daughter of the lemon tree. But instead of the beautiful girl he remembered, he found himself looking at a dark-faced woman who insisted that she was his wife and that her skin had merely taken on a brown tint from all of the gall that had coursed through her veins in his absence. The prince sensed that his tower now harbored a terrible secret, and sadness dug its claws into his heart.

Though the golden-haired woman with the brown skin tried to be attractive and to win the prince's love in a thousand ways, she did not succeed. Once when he was sitting at the tower window where the golden-haired daughter of the lemon tree had spent so many sad days, he noticed a goldfish swimming about in the pond below with such grace that it filled his heart with joy just to look at it. From then on, he spent most of his days watching the goldfish, until the crafty maid realized that it was none other than the golden-haired girl herself swimming about in the muddy water. And so she set about once again to put an end to the daughter of the lemon tree.

One day, pretending to be ill, she asked the prince to serve her the goldfish for dinner, saying that perhaps this remedy would restore her former beauty. Dutifully, the prince ordered the fish to be caught, cooked, and served to the woman on her sickbed. The dark-skinned girl gulped down the fish greedily, hoping in this way to put a stop to the golden-haired girl and at the same time cover up her own wretched deed forever. But her mistake was to throw the fish's bones out the window. And that very evening there sprouted on the same spot a eucalyptus tree whose blossoms reached up to the window in which the prince sat, and leaned towards him as if to look into his face.

The prince felt the same love and affection for the eucalyptus tree as he had felt for the goldfish. This did not escape the dark-skinned girl's notice, and she ordered the tree cut down. But as the gardener raised his axe, he heard a voice say: "Strike with care, and make sure that you don't cut into me!" The gardener struck the trunk of the tree very gently with his axe, listening all the while to the voice, which now said, "Be careful, for here within is your Lady. . ." When he had reached the middle of the trunk, he realized that it was hollow, and out came Goldenhair in all of her beauty, begging the gardener not to say anything to anyone, but to hide her instead in his house for a short while, which he did.

The day that the tree was cut down, the prince fell ill, and lay for weeks in his sickbed, refusing to eat or drink. Goldenhair cooked him a fine chicken soup and had the gardener bring it to him. The faithful gardener tried to get the prince to eat the soup, but he refused. The gardener insisted, holding the spoon up to the prince's lips, hoping he would at least try one spoonful. Finally the prince gave in, and while sipping, saw Goldenhair's ring in the soup. "This is the ring of my Goldenhair," he cried out. "Where did this come from?" The righteous gardener told him the whole truth, and the prince's good spirits were immediately restored.

Then the prince proceeded to the gardener's house, where he found his Goldenhair waiting for him. After holding her and kissing her for a long time, he asked her to tell him all that had happened. When he heard the truth, he summoned his counsel of twelve to call a great crime to trial. When the people had all gathered around he rose and spoke, "Once there was a gardener who had an apple tree with unusually good apples, which stirred the gardener's envy. Instead of picking the fruit, he cut down the entire tree. I bid each one of you to tell me what he thinks about this gardener."

The first one to stand up was the untrue and vainglorious maidservant, who didn't realize that the story was about her. "This deed is worthy of death," she said, and recommended that the perpetrator be tied to two wild horses that would be sent in opposite directions.

The prince then rose and told the people about the maid's wicked behavior. Their consensus was that she should be put to death in the manner she herself had described. So the servant girl, who knew nothing of gratitude or grace, found a fitting end. The prince and Goldenhair spent the rest of their days in joy and happiness.

This folktale[27] comes from the Greek island of Rhodes. It starts off with a description of a happy royal family: a king and queen who are known far and wide. They have only one child, but he is extremely beautiful. Problems do not arise until the prince turns sixteen and starts playing with the vizier's son, whom he loves as his own brother. Completely by accident, he throws a stone down a chimney, smashing an old woman's cooking vessel, in return for which the old lady puts him under a spell that grips him with desire for the daughter of the lemon tree. The prince begins to fret, stops eating and sleeping, and becomes ill.

This is the initial situation of the folktale. Apparently the family was happy until their son started playing with the son of the vizier. Out of this a story can develop. The two friends throw stones. What seems to be a release of harmless aggression has the unfortunate consequence of smashing a pot on a poor old lady's hearth. This image of a pot in which lentils have been boiling for some time gives us a sense of what lies beneath the royal family's superficial happiness. The hearth is the heart of the household, the place where meals are prepared and trans- formative processes take place. It is a maternal realm, expressed by the old woman cooking in her kettle. The stone falls directly into this mother complex. Throughout the story, the prince's mother is described as careful and protective—somewhat like the pot, which contains things and keeps them together. Thus we may conclude that the pot on the lower level mirrors the family situation above: This family is like a "pot" in which everyone stays together as long as possible and everything is kept in the best of order. That the parents intend to protect their son from life is suggested in their promise to obtain the princess for him right away—to satisfy his every wish immediately. I feel that this describes a family system characterized by over- protectiveness. Indeed, the fact that the prince only began to play with the vizier's son at the age of sixteen suggests that his parents tried to keep him in the "pot" as long as they possibly could. The vizier's son is a brother figure of the kind that emerges when a step forward in consciousness is called for. He seems more capable than the prince of playing out his harmless aggression, which of course results in the pot going to pieces.

How does the prince react to the old woman's curse? He immediately gets depressed, frets, stops eating and sleeping. He does not dare tell his parents what happened, and indeed he himself knows only that he has to go and find something, though he has no idea what it is or where to find it. Here we see how the symbiotic family system makes one susceptible to depression whenever change is demanded. But the prince is not susceptible to depression alone; he is just as easily made happy —as evidenced by his immediate recovery when his parents, in their overprotective style, promise to get him the daughter of the lemon tree. But the prince is not completely without his own ideas: He decides to visit the old lady to ask her forgiveness and her counsel as to how to find his heart's desire.

But the issue here is not simply finding the daughter of the lemon tree. In her curse, the old woman prescribes a particular quality and pattern for the prince's relationship with this feminine figure. He should crave her every bit as much as she craves her ruined dinner of lentils; he should be driven by a longing for something that can never be obtained. And when he does find the daughter of the lemon tree, he should be stirred by the wish to "eat" her, just as the woman wanted to eat her lentils.

This is a very common sort of symbiotic situation: Out of an overindulgent family system, there grows the image of a prince or princess that "someday will come" to fulfill the urge to merge. This longing, which is impossible to quench in reality, may be transferred onto areas of life where there is a chance it may be fulfilled. If there seems to be no chance for fulfillment, depression sets in. If fulfillment does seem possible, mild mania may result.

The old woman's curse indicates how completely the prince is dominated by the mother complex, whose present focus is keeping and eating (and longing for something impossible). The spoiled prince must wander eastward by himself for three years, wearing out three pairs of iron shoes. He must stick to his path and may not stop to rest. After the third year, the old woman tells him, he will find the daughter of the lemon tree.

For this honest (if not very encouraging) information, the prince enthusiastically thanks the old woman. Then he immediately goes to tell his parents what he is about to do. Naturally they are afraid of losing him and try to talk him out of it, but in the end they grant their permission—for fear that, if they don't, he will fall ill again. Once again, we see evidence of their over-protectiveness in the black attire they wear in contrast to the prince's elevated mood. They fear the worst and mourn in advance. Of course, they really do have occasion to mourn, since the family pot has indeed been broken. On the other hand, the prince's elevated mood at the beginning of his journey—his denial of the three years that lie ahead, during which he will be wearing iron shoes—shows how out of touch he is with the gravity of his situation. Thus we see the symbiotic group's tendency toward either mania or depression. What is the meaning of the prince's iron shoes? They are heavy, weigh him down, even drag him to the ground. As such, they seem a fitting cure for someone who dances on rooftops and casts a casual glance at reality. Iron is the metal of Mars, the aggressive god of war. A bit of aggression, a bit of iron will, seems to be required to sustain a journey like the one the prince is about to make, during which he will not be able to rely on anyone's help. The shoes are an emblem of this call to become more determined and persevering; he must resist the temptation to flinch from reality or be swayed from his intentions, facing always toward the east and the rising sun.

This is the first step out of overprotectedness, and is a requirement for anyone who comes from such a family stewpot. A longing for the near-impossible drives the prince on a path that turns out to be, for the most part, quite grim. This is the way out of "the easy life." He cannot allow himself to rest, for in resting he would be vulnerable to regression. The path he takes is very much in line with his temperament: For a while it ascends steeply, then it descends again just as precipitously. And yet he does not digress. Here we see the strength of this family vessel: He does what the old woman says without asking too many questions; he sees the matter through.

But his perseverance does not last forever; when the sun sets after the third year, his spirits sink and he begins to lose faith. Just then he sees the castle, his mood lifts, and he speeds (even in iron shoes!) to the "palace of destiny." Not only is his personal fate decided in this castle, but so is that of humankind. Strangely, the workers here neither look at nor speak with one another. If this is the palace of destiny, fate must be oddly disconnected. (Contrast this with the prince's intense connectedness at home.) And when the prince is ignored—like everyone else here—he loses faith again. The family pot out of which he grew provided him with a great deal of attention, and so to be deprived of it now is painful.

But finally, he is noticed. A voice from upstairs beckons him to come closer. He meets a man the same age as himself, whose name is, in essence, "Your Fate." We are reminded here that this is a Greek tale. At least as far back as Plato's dialogues, the figure of the *daimon* played an important role.[28] A personal guardian and guide of souls, this spirit of destiny embodies the law of one's own nature. At the same time, it belongs to the world soul, and therefore knows the future. Like a *daimon*, the spirit of fate in our tale dwells in a realm where fate is decided and appears to know the future, for he promises the prince that his longing for the daughter of the lemon tree will be satisfied.

Thus we see that the prince's long and lonely road leads him to an experience of his own, entirely personal fate—to the crux of what he must do, which has nothing to do with what others want him to do. He must discover *his* fate, and indeed he does. I see the prince's relationship with the spirit of his own destiny as bringing him yet further out of symbiosis. And yet even this spirit seems somehow under the influence of the mother complex, addressing the prince as "my child." The spirit seems to have nothing better to do than provide the prince with food and a bed. Moreover, he is the same age as the prince! And so we might ask how his three years of wandering have helped the prince work through anything associated with the mother-complex.

As for the prince, he does not care to eat or sleep. He only wants the daughter of the lemon tree, and he wants her now! A step forward is in the making.

The spirit of destiny then claps his hands, and a winged horse with a human voice appears. Of course, we know of such a horse from Greek mythology: Pegasus, born of Poseidon and Medusa. Poseidon was the god of the sea, who was called the "earthshaker" when he appeared with his horses. He embodies a highly vigorous aspect of the psyche, in contrast to Medusa, who petrified anyone who looked at her. Pegasus thus represents a combination of an extremely powerful drive and an equally strong inhibition, a dynamic that is associated for the most part with the sky and spirit. In one tale, Pegasus stamps a well out of the ground that becomes sacred to the Muses. Thus, for many poets, Pegasus has been a symbol of creative, dynamic energy. We say that we are "on the wings of Pegasus" when we are seized by creative imagination—a power of imagination propelled by the power of instinct. Whenever we are seized by instincts that are blocked from actualizing themselves, we tend to sit "on the wings of Pegasus" in order to at least realize ourselves in imagination. This can be positive, as in an experimental or artistic activity, or negative, as in a flight from reality. Whether fantasy proves to be constraining or enriching depends on how it is translated into reality and integrated into everyday life. At this point in the story, the prince appears to me to be seized by the creative imagination. He is instructed to sit on the horse and blindly obey whatever the horse tells him to do. We already know from the beginning of his journey that the prince has blind faith both in himself and in others as long as there is a way out. When he no longer sees a way out, his trust crumbles—which is both the benefit and the liability of an overprotective, symbiotic family system. And now he takes off on a really high flight. If this had been one of Plato's dialogues, we would have met with the spirit of destiny in these "celestial regions." His magnificent flight conveys the prince to a garden with a tree bearing three lemons and guarded by forty giants. In every lemon resides the daughter of the lemon tree. All are guarded by the giants, who apparently do not see what comes down from above. By swooping down on the winged horse, the prince is able to grab one of the lemons. The tree-mother screams and the giants run around in a frenzy, but they see nothing. The prince has his lemon, and the spirit of destiny tells him to cut it in two in order let out a beautiful girl with golden hair.

The garden the winged horse brings the prince to is reminiscent of paradise, as well as the place where Hercules stole the golden apples of the Hesperides, which granted immortality. The woman is a fruit who must be torn away from her mother, the lemon tree, a separation that provokes desperate tears from the latter. Even here, so far from the original royal court, the mothers seem to want to hold on to their children. We can see from the number of guards posted around the tree just how precious this daughter is. Here is another image of the intense protection that symbiosis offers. Nothing should be allowed to enter from outside. Tree and fruit are often characterized as mother and child. In German we say that "an apple falls close to the tree" and that children are "little fruits." The apple tree plays the same role in our cultural symbolism as the lemon tree does in Greece. The fact that the prince beholds his future wife in the form of a fruit may be the result of the old woman's curse—i.e., his mother complex—in which containing, nourishing, and oral gratification were prominent.

In order to save themselves three years of travel, the couple sets off for home on the winged horse. Their wedding is celebrated with festive jubilation. This entire passage, which deals with the abduction of the lemon tree daughter, can be understood as the prince's creative imagination. In his fantasy, he makes his spouse into an extremely important and precious creature who dwells in the realm of luminaries. Sparked by his longings, his imagination may have created an image of woman that is exaggerated beyond all proportions. The woman who comes out of a lemon has no past; she exists for him alone. He produces her magically and brings her to life. This fantasy of a woman does not necessarily preclude the existence of a real woman, but if there was a real woman, she would have to swallow the fact that her husband had married an image rather than her. Despite this slight hitch, the prince's path away from home—which has so far led him to the spirit of destiny and the lemon tree—has now led him to develop and adore an inner image of a young woman; he could, after all, have stayed at home with mother.

But the prince's happiness does not last long. If it did, things would have become much too symbiotic again. A neighboring king declares war on the prince—a conflict arises that demands encounter. The conflict originates with a neighboring king; that is, the prince experiences aggression as coming from the outside. Now his own aggression is called into action, along with his will to fight. Wars in general are the sign of a failure to reach an agreement concerning disputed claims, leaving violence as the only alternative. But here war also implies that it is necessary that the prince depart in order for something new to happen. The theme of separation comes up alongside the need to fight something in the outside world. Here we see yet another feature of the way out of symbiosis. When someone is tightly wrapped in symbiosis (especially when the relationship seems to an onlooker particularly happy and harmonious) aggression is sparked among those who have no part in it. Feeling excluded, they strike first, providing those involved in the symbiosis with an opportunity to resolve their symbiosis. However, the attack can also have the opposite effect, driving the other party yet deeper into symbiosis. But this is not the case with the kind of symbiosis that our tale deals with, which is filled with energy and serves as a source of nourishment. Thus it has a propensity to maintain itself. People with this kind of symbiotic tendency often have the energy to deal with conflict, even if they do not particularly like to. This is suggested in our tale in the description of the prince's guardian spirit as a "strong young man."

Before going to war, the prince builds a tall, sturdy tower to enclose Goldenhair and her maid, and sets up a number of guards around it. The image reminds us of the giants stationed around the lemon tree as well as the lentils safely held within the pot on the hearth at the beginning of the tale. The tower shows how possessive the prince feels about Goldenhair— though no doubt he would say he was only concerned with protecting her. Once again, we see the desire to conserve causing stagnation and keeping Goldenhair apart from what life has to offer. No wonder she first falls into a depression and then into a deep slumber. Completely insulated from life, all she can do is regress. Yet she is not completely alone; the dark-skinned maid is there with her. Just as the prince has to fight with an aggressive king, Goldenhair has to contend with the dark maid—which is to say, they both constellate their opposite, dark sides. Until now, it seemed as if everything transpired in a sphere of light, beauty, and harmony. Now the wicked maid casts Goldenhair into a "deep pond full of muddy water," an image of stagnation, regression, and depression. Whereas previously she had been "up" in a lemon tree , now she is down in the dumps. All of the prince's attempts to protect and preserve Goldenhair have been to no avail; indeed, they have caused her disappearance at the hands of the dark-skinned maid.

The foregoing sequence can be viewed from various perspectives. Symbiotically-bound men, such as the one in this tale, tend to create an idealized image of woman—suggested here by the girl's golden hair and the "celestial realm" from which she comes. If such an image is actually found in reality, it surely must not be allowed to change. In an effort to prevent any doubts, great measures are undertaken to protect it from being spoiled by outside influences. The image is not only protected from outer things; it is sealed off from the self as well, like a photograph that one carries around of a moment that has been trapped and forbidden to change. When the quality of the actual experience vanishes, the original emotion floats away, leaving intense sadness. One thinks that the emotion is still there—the dark maid puts on the lost girl's golden hair—but somewhere within one knows that it is dead and gone.

From another perspective, we see a man who has imagined a woman to be the way he wants her to be. He married an image rather than a person, and he will not allow this image to "come down" to reality; the tower preserves her perfection. In comparison to such a fairy-tale image as we have here in Goldenhair, every real woman must seem like a "dark maid," fraught with characteristics that are not light and wonderful. This reality slowly makes the vision of the ideal woman disappear, replacing it with a deep sadness. I think for example of a man who wants his wife to provide him with constant inspiration, to help him with his career and do so with lots of affection and good cheer. In a moment of resignation, he must admit that his wife cannot live up to his expectation. She hardly ever inspires him; she criticizes him when he complains, she is often boring, and so on.

This state of affairs is reflected in our story at the point where Goldenhair exists no more; the inspiring fascination is gone, and *sadness* must be tolerated. Explaining that her dark skin tone was caused by the excretion of gall that she experienced in the prince's absence, the dark maid links the symbol of the tower with the emotion of sadness. Gall is often associated with pent-up rage and spite in relationships; as "black gall," it is associated with melancholy. But now it is not only the feminine which carries this emotion; sadness "digs its claws" into the prince's heart as well. The prince does not run away from his sadness; he seems to take up residence in the tower and, rather than becoming ill or living out his depression somatically, becomes sad as he faces the depressing situation directly.

When the golden fish starts to move in the muddy pond, the prince sees a flicker of promise in the midst of his muddy dilemma and feels a glimmer of the tremendous feeling he had for Goldenhair. It may simply be a glimmer of hope that what he has lost might come back to life in some form. Seeing through the darkness, his eyes are newly opened (on the "ground" and in the depths) to the object of his longing.

But from her dark tower, the evil maid still reigns. She will not allow this existential feeling to emerge, and so devours the fish immediately. And yet by doing so, she gives the goldfish the chance to reappear in another form—as a eucalyptus tree. A tree is something that grows over a span of years. Connecting the ground with the heights, it doesn't move, but remains stable; in contrast to the fish, one can hang on to it. It is also physically closer to the prince than the fish; we are told that its blossoms reach nearly into his window.

Knowing that Goldenhair has come from the fruit on a tree once before, we suspect that she will soon emerge out of her regression, that the existential feeling that came to the prince through Goldenhair will soon come alive again, only on a deeper level this time, having undergone transformation. But this existential feeling must be protected from the maid, who wants to have the tree cut down—and protected as well from the part of the prince that no longer wants the feeling to come up, no longer wants to lose himself in large ideals for fear of being disappointed. And once again, the dark maid ends up facilitating further growth: the gardener cuts the tree down, and out of the trunk steps Goldenhair.

The motif of the woman who emerges from a hollow tree— a birth out of a tree, so to speak—is a common one in folktales, representing resurrection through the mother. It is hardly surprising that this motif would emerge in this context, considering how strongly the tale is colored by the mother archetype. After all, the girl was clearly stolen from her mother (the lemon tree); it is only logical that a regression would have to take place following this growth away from the mother. After overcoming all that encompasses and encloses—garden, tower, water, and finally tree trunk—Goldenhair should now be prepared to venture out into life and reality; she no longer needs to be enclosed or in-teriorized.

In terms of the prince, or of a man's psyche, this development suggests that he has enlarged his capacity for dealing with women, feelings, and reality in general. From the perspective of a feminine psyche, we could see Goldenhair as a woman who was symbiotically bound to her mother before she was "stolen" by a man who then sheltered her from life. She sinks into boredom and depression, to the rock-bottom of her being, and slowly comes back to life as a fish and a tree, out of which she is born again.

The figure of the gardener is also noteworthy in this regard. Gardeners are primarily caretakers rather than consumers. In the appearance of this figure at this point in the story, we can see that a new side of the prince has emerged.

But I am jumping ahead to the happy ending. First the prince must become ill and lie in bed. He gives up on everything—lapsing into a condition familiar to any therapist of depressed clients: A development prepares itself in the unconscious, becomes visible and tangible (without excluding opposing tendencies), and the moment the breakthrough is about to happen, the client collapses into hopelessness.

In our tale, Goldenhair now becomes active for the first time, showing how worldly she can be. She cooks a soup for the ailing prince into which she puts her ring. The ring is a symbol of relationship, of being bound and joined, of fidelity throughout all transformations, and thus also a symbol of wholeness. Thus we can say that the story's final step, a new bond between the prince and a transformed Goldenhair, was not his idea but hers. His part was to endure the transformations, to survive the dark maid without ever completely losing hope that Goldenhair would return. Now the relationship can take place on a different level, a level less influenced by the mother complex and thus more realistic. The prince has now acquired a capacity for relationship, something that only occurs when one can see and tolerate the very real possibility that one may lose one's companion. No longer does he require the maid, who, as Goldenhair's opposite, had the function of disturbing and eliminating her glossy one-sidedness. Perhaps, too, he has lost his propensity for falling into extremes, an attitude of "either everything is good or everything is bad." Now at least one of those extremes eliminates itself—the dark maid. In light of this, it is interesting that the maid is so often described as "unfaithful," since the tale portrays the tendency to fall into extremes as unfaithfulness and the prince really is keeping "faithfully" to his path.

Let us attempt to visualize once more the path out of symbiosis depicted in this tale. We are presented here with a symbiosis in which the mother complex plays a large part, providing a sense of bounty, nourishment, protection, and shelter, but also creating problems of overprotectedness, lack of reality-orientation, and unreasonable demands; the typical reaction to problems in this symbiosis is depression.

The passage out of depression begins when the prince throws stones, that is, when a natural aggression develops that gets him out of the house. There follows a phase of endurance, of trudging through life with no regression permitted. The iron shoes make him conscious of every step. The goal of this phase is the discovery of the spirit of his own destiny, which we interpreted to mean his fate independent of what others might think. This is why he has had to remain alone for three years, plagued by a yearning he cannot satisfy.

And now he has the grand vision of Goldenhair. It is his task to discover the feminine, but the searching and winning are still tainted by the original mother complex. He is elevated, dreaming up a lofty idea of the feminine that he can win for himself by storm. Then comes another symbiotic phase, which ends when the problem of aggression becomes acute again and when the desire to preserve the status quo (depicted in Goldenhair's tower imprisonment) becomes so strong that only stagnation and depression can ensue. This depression is absolutely necessary; it drags the prince into his own depths and dashes to bits the lifeless image of the woman he had painted for himself.

The passage out of depression winds through the territory of mourning, as the prince mourns the loss of Goldenhair. Because he sticks with his grief, she comes back to him out of the dark pond—in another form. Thus we see that enduring the pain of loss leads to the transformation of both partners and to the possibility of relationship.

■ Jorinda and Joringel

There was once an old castle in the middle of a large, thick forest. In the castle, there lived an old woman who was an arch-sorceress. During the day she changed herself into a cat or a night owl; at night she returned to her human form. She would lure wild beasts and birds into her snares and then kill them and boil or roast them. If anyone came within a hundred yards of the castle, they would be frozen in their tracks until she said the magic words that would release them. And if a virgin came within her domain, she would turn her into a bird that she would trap in a basket and take into a certain room. She must have had seven thousand such baskets in her castle, filled with all of the rarest birds.

There was once a girl named Jorinda who was more beautiful than all the others. There was an even more beautiful boy named Joringel whom she had promised to marry. They were engaged and spent their days in pleasure and joy. Once they went for a walk in the woods, in order to talk without being disturbed. "Take care," said Joringel to Jorinda. "Don't get too close to that castle." It was a pleasant evening. The sun cast shafts of green light between the trunks of the trees. From an old birch, a turtledove sang its song of woe.

Suddenly Jorinda felt inexplicably smitten with sadness and began crying. The tears came and vanished and returned again. As he sat down in the sunshine, the same thing happened to Joringel. The two of them felt as forlorn as if they had been awaiting their deaths. Looking all around, they lost their grip on themselves and could not find their way home. The sun was halfway up and halfway down. Through the bushes, Joringel saw the old wall of the castle nearby, which startled and frightened him to death. Jorinda began singing:

> "My little bird with the red, red ring,
> Sing sadly, sing sadly, sing sadly.
> Sing to the turtledove of your death,
> Sing sadly—coo-coo, coo-coo."

Joringel looked at Jorinda, who had changed into a nightingale as she sang "Coo-coo, coo-coo." A night owl with glowing eyes then circled around her three times, letting out a screech with each swoop: "Shoo, hoo, hoo, hoo." Joringel could not move. He stood there like a stone, unable to speak or even cry. Hand and foot seemed frozen stiff. After the sun set the owl flew into a bush, out of which then came a skinny, old, bent-over woman, with yellowish skin, large red eyes, and a twisted nose whose tip reached her chin. Muttering something, she caught the nightingale and carried it away. Joringel could neither say anything nor move from the spot; the nightingale was simply gone. After a while the woman returned. "Greetings, Zachiel," she said in a muted voice. "When the moon shines in the basket, let him go, Zachiel, for better hours." And so Joringel was released. He fell on his knees before the woman and pleaded with her to give him back his Jorinda. But she swore he would never have her back again and vanished. Though he called out to her, moaned and wailed, it was of no use. "What will become of me?" he cried, and sadly went on his way.

After a time, Joringel came to an unknown village. There he stayed and tended sheep for several years. Often he would circle around behind the castle, but he always took care not to get too close. Finally one night he dreamed that he had found a blood-red flower, in the middle of which was a large, beautiful pearl. He picked the flower and took it with him into the castle. Everything that he touched with the flower was freed from the witch's curse. With this magic flower, he even won his beloved Jorinda back.

In the morning when he awoke, he began searching every hill and valley for such a flower. On the ninth day of his search, he found, early in the morning, the very same blood-red flower that he had seen in his dream. In the middle was a drop of dew as large as the most beautiful pearl. He carried this flower with him day and night until he reached the castle. This time, when he got within a hundred yards he was not frozen but was able to keep going until he reached the castle gate. When he touched the portal with the flower and it sprang open, Joringel was overjoyed. He then proceeded through the courtyard, where he heard the sound of many birds. He went on un-til he came to the room where the sorceress was at that moment feeding the seven thousand birds in their wicker cages. When the sorceress saw Joringel, her anger flared up. She berated and scolded him as if she were spitting gall and poison. But she could not pene-trate the circle that surrounded him for two yards on all sides. So Joringel passed the sorceress and walked into the middle of the room filled with the birds in their baskets. There were many hundreds of nightingales and there seemed no hope of ever finding his Jorinda among them. But then Joringel noticed that the old woman had secretly removed one basket and was start-ing with it toward the door. Leaping after her, he touched first the basket with the flower and then the old woman. This deprived the sorceress of her magic and caused Jorinda to appear on the spot. Jorinda, as beautiful as ever, threw her arms around Joringel. He then used the flower to turn all the other birds back into girls, went home with his Jorinda, and lived happily with her for a long, long time.

––––––––

This is a German folktale,[29] recorded during the Romantic period by the Brothers Grimm. The title itself hints at a prob-lem: the names Jorinda and Joringel are practically identical. This could suggest two aspects of one person, masculine and feminine, but it might also mean that the two characters have such a close relationship that their differences have been erased—in short, they have a symbiotic relationship. The tale begins with an arch-sorceress who lives in an old castle in the middle of a large, thick forest, who changes herself by day into a cat or an owl but who is a person at night. Those who come too close to her are paralyzed. She changes pure young girls into birds, whom she encloses in a basket and then takes into a room where she keeps them as prisoners. The girls she catches become more and more entrapped—first as birds, then in baskets, then in a room—as if ever more veils were being cast over them.

Turning first to the arch-sorceress, we note that she portrays something that has been repressed. Terribly alone in the thick of a dense forest, she is familiar with animals. The animals into which she changes by day—the cat and the owl—closely describe her nature. For the cat, a parallel from mythology is provided by Bast and Sekhmet, two ancient Egyptian mother goddesses who were depicted with the heads of cats. Bast was the good cat, Sakhmet was the enraged one, who was often portrayed with the head of a lion as well. It makes sense that cats would have to do with the feminine: whenever the expression "cat" is used in referring to a woman, one is speaking from an erotic perspective. It seems to make sense as well that there would be one goddess for the good cat and another for the bad one. Anyone who has had a cat as a pet knows how soft their paws are until they decide to use their claws. Cats symbolize an instinctual femininity; they like to be cuddled, but they are also independent and unpredictable.

In ancient Greek mythology, the owl was the bird of Athena, the goddess of wisdom, war, and handicrafts. The owl symbolizes nocturnal, prophetic, and intuitive wisdom.

One could say that the tale's initial situation depicts a historical situation in which the instinctive-feminine and the spiritual-feminine (the cat and the owl) are repressed, creating fears and anxieties about falling under a curse. Yet the magical and prophetic dimensions of life are not only repressed but sought after. One not only fears being cursed, but longs to be seized and inspired. Here, we are dealing with a folktale of the Romantic era. Coming after the Enlightenment, in which feeling was "banned" and relegated to a "deep forest," the culture of Romanticism was one of unparalleled feeling.

Joringel knows that it is wise to watch out for the castle and not approach it too closely, and yet the castle exudes a dangerous attraction. Considering the seven thousand baskets with young birds in the sorceress's keeping, her curse must be very powerful. It would seem that falling in love exposes one to extreme danger, as in the case of Jorinda and Joringel. Love really does hold one spellbound, and here this magic is hardly in the service of life. Both Jorinda and Joringel have a strong foretaste of what is to come and are as sad as if they were visiting their own graves. The sun sets on this scene, bringing on the night.

When two people are very much in love, not only is the mother complex constellated but so is whatever is socially and historically problematic about the mother complex. Jorinda's and Joringel's symbiotic relationship leads "straight" to the arch-sorceress, to a realm in which the man is completely paralyzed, rigid, expressionless, and unable to communicate with his woman. In this same realm, the girl becomes a nightingale—a bird whose song is supposed to be sorrowful and full of longing but also seductive and stimulating. But as a nightingale, she cannot be reached, and the possibility of relating to one another is lost. The sorceress is the one responsible for the woman's entrapment as a nightingale.

Symbiotic love between two people can be a real trap. Intensely desirous of love's magic (in this case, fueled by a historical movement), a couple in love elevates the woman and fashions her into a nightingale. Deprived of her human voice—and of communication—she becomes both super- and subhuman. Meanwhile, the man is petrified, incapable of acting or of winning back his Jorinda. The relationship is severed.

In another respect, the seven thousand nightingales in this story can be seen as the many "beautiful souls" who have lost interest in real life, like the many dreamers who fell prey to Romanticism's beckoning call and were pulled away from reality.[30]

Symbiosis yields to a separation. By means of another magical saying, Joringel is released from his petrification. In this version of the tale, the sorceress is interested only in girls. However, in another version, it is the girl who is forced to tend sheep while the young man is taken into the sorceress's keeping.

What does it mean that Joringel goes to tend sheep? To begin with, this is an activity he undertakes on his own. He must accept the separation. Tending involves keeping the flock together, keeping things from becoming scattered. The protagonists of folktales who tend sheep are tending to themselves, gathering their vital energies. This self-assembling is also expressed in the image of Joringel circling around behind the witch's castle. He seems to move within the problem's circumference without getting too close to it. Tending is also an act of introversion; he reflects on himself and so must endure intense grief.

Finally one night he has a liberating dream. He dreams that he finds a blood-red flower with a large, beautiful pearl in the center. The flower gives him the power to undo the curse, an image that fittingly describes a therapeutic process. Joringel surveys the problem from every angle. He engages himself in a process of collecting his energies, attending to himself and his vision. One day he has a dream that suggests a solution to his problem, and which emphasizes the value of the flower's center.

What sort of power might the blood-red flower symbolize? Red suggests passion and suffering, blood, and corporality. Flowers often symbolize feelings and eros. The red flower suggests the passionate feeling of love, including its physical aspects, and yet here it is related to the pure white pearl as well. The blood-red flower also indicates a connection with Jorinda, whose song to the turtledove speaks of a bird with a small red ring.

The pearl implies great preciousness, something that has been brought to perfection. Among mystics, it is the symbol of enlightenment, the unity of the divine and human realms. Growing concentrically, the pearl symbolizes an enlightenment that unfolds gradually. Thus the growth of the pearl can be see in relation to Joringel's circular path around the castle. To my mind, the connection between the red flower and the white pearl depicts a union of carnal and mystical love, which has fallen within the folktale protagonist's realm of experience. Love's infatuation no longer need paralyze him; having discovered and experienced it within himself, he can now ward off the witch's power over him.

The sorceress undid the curse on Joringel by speaking the mysterious phrase "When the moon shines in the basket, let him go, Zachiel, for better hours." Though it is not clear what this verse means, we can note a structural parallel between the moon in the basket and the pearl in the blood-red flower. Perhaps, then, the marriage of opposites must take place within Joringel himself before he can rediscover Jorinda.

Having learned how to "see," Joringel goes immediately to find in reality what he has seen in the dream. And indeed he does find a flower which contains a drop of dew, the sign that morning has come and the night of suffering is over. That he recognizes the pearl in the drop of dew means that he can see through presented reality into its transcendent background. Now that he has integrated within himself what the arch-sorceress previously embodied for him, she no longer has power over him. Even more, having discovered his middle ground, he is centered within himself and thus able to rescue Jorinda. Having undergone a mystical experience, he no longer needs her to be the nightingale for him. A real relationship can begin. The sorceress has lost her power over him, as well as over Jorinda and the seven thousand other birds.

The tale describes a symbiotic relationship between two lovers. Of course, being in love is always symbiotic to a degree, but the similarity of the partners' names alerts us to the fact that the symbiosis here is particularly strong. The story springs from an historical environment in which the instinctive-feminine and feminine wisdom, along with nature mysticism, reside deep in the forest, but exercise a strong attraction nonetheless. Succumbing to this attraction would mean transferring onto woman this entire sphere of feelings through dreams, expectations, and anxieties. When the woman triggers such a storm of feelings, relationship is no longer possible; the man is petrified and the woman becomes a nightingale—that is, she is robbed of her human form and can no longer react as a person.

Joringel must find the way out of symbiosis. Although he has been cursed by the sorceress, he has not been taken prisoner, and is capable of waking up. He must survive his separation, gather together his energies, and reflect on the problem of loss and on the arch-sorceress who caused this loss. He must ruminate on what has caused this curse and robbed him of his autonomy. But he must also think of his Jorinda whom he loved so much.

Turning inward in solitude, he receives a dream that undoes the curse by expressing the intense feeling that he has discovered within himself: the union of physical and spiritual love. His symbiotic needs are thus translated into an inner experience of transcendence. What was formerly expressed as the cat and the owl on an unconscious level is now available as a conscious experience. As this feeling grows within Joringel, the sorceress loses power over both him and Jorinda.

This folktale proposes that in a symbiotic relationship, the partner who is less paralyzed by the symbiosis must be the one to take the first step toward creating the conditions under which a new situation can develop.

■ Concluding Remarks

I find it striking that in all the folktales we have discussed, symbols of enclosure figure so prominently. In "The Wife," it was the grave; in "Journey to the Underworld," Sina's hellish whirlpool, Faalataitauana's creepers, the shells containing the three brothers, and the many instances of sleep; in "Redhair Greeneyes," the son's initial passivity; in "The Daughter of the Lemon Tree," the several variations on the symbol of the "pot," the paradise guarded by giants, and the tower, fish and tree; and finally, in "Jorinda and Joringel," the seven thousand birds entrapped in baskets.

Such symbols of enclosure, entrapment, protection, guarding, and safekeeping appear in most folktales, but their preponderance in these tales of symbiosis is conspicuous. Tales in which the mother is missing and the son must set out to find a woman, may have many parallels to these stories, but there, the motif of enclosure does not occur with such regularity. For images of entrapment and containment are associated with the mother archetype. Thus, to judge by these tales and their imagery, we would have to answer affirmatively the question we posed at the beginning of this section: namely, whether or not there is a mother problem behind every symbiotic relationship (keeping in mind, of course, that mother problems come in many sizes and shapes, and may have a variety of consequences for the individual).

We can see what is characteristic about symbiosis by comparing the folktale heroes we have been studying with those whose mother issues are less pronounced. For one thing, the latter have a greater capacity for aggression. For another, they find the way into life more attractive than the way out, in spite of their bonds to mother. For the folktale protagonist in a highly symbiotic situation, the decision of whether to live or die is not at all easy, and creates the tale's drama.

In the four folktales discussed here, I have attempted to describe types of symbiosis found in various life situations and on different levels of consciousness. What all these tales of symbiosis share is the symbol of enclosure, indicating the latent presence of a mother complex oriented toward keeping and containing—and, naturally, nourishing as well. Also typical of these tales is the way autonomy is short-circuited and aggression is absent or blunted, all of which creates a characteristic problem in relationships.

Common features have emerged among the "ways out of symbiosis" we have discussed, as well. These commonalities can only be described very generally, since in the final analysis, every form of symbiosis must be considered individually. Still, we can observe that whenever symbiosis goes too far, it leads to an expulsion. Forced separations may lead to change, but they can also result in deep regression and even, in some cases, death.

In dealing with symbiosis, I find it helpful to distinguish between creative and regressive confrontation. Admittedly, there are regressive elements involved in nearly every creative confrontation, although there may not always be creative elements in every regressive reaction (witness, for example "The Wife").

It is hard to predict which factors in a given folktale are likely to result in the choice to deal creatively with symbiosis, and which would result in the choice to deal regressively. It seems to me that this will depend on which quality prevailed during the initial phase of the symbiotic bonding. Was it primarily a matter of nurturing or of preserving? Was it a nourishing sort of keeping, or one that concealed aggression? Among the various

paths to change, the path of separation from the symbiotic partner continually presents itself as an option. Separation usually
involves a change of location and of behavior. New ways of acting may include a greater readiness to take risks, to confront
people, and to mobilize one's aggression. Following the initial
separation, the symbiosis is generally transferred to something
or someone else, leading to a new separation and often a resulting spell of grief-work. Such a path leads to a new starting point
where the protagonist is usually more autonomous, if not yet
entirely free of symbiotic tendencies.

Our folktales also suggest that the journey out of symbiosis
prepares the individual to reenter relationships in a renewed
way. Though the protagonist may not have overcome all symbiotic tendencies, he or she has become aware of a rhythm,
realizing that a phase of separation and individuation must follow every phase of symbiosis.

It has been my aim in this book to examine various types of
symbiosis and various ways out of them, in hopes that individuals in therapy might be thereby encouraged or inspired. I
would like to conclude by attempting to summarize these.

In "A Journey to the Underworld through the Hellish
Whirlpool of Fafá," we started with an incestuous family characterized by a high degree of unconsciousness. A mother's
unconsciousness, for example, makes it difficult for a child to
get enough distance to make her own choices and discover her
own identity. Such a child may develop an overactive unconscious and may fail to develop an ego sufficient to deal with it,
as is often the case with people with chemical dependencies.
One aim of therapy in these cases may be to reinforce the nurturing aspect of the mother archetype. Yet it would be crucial
for the client also to practice autonomy at every opportunity in
order to reduce the likelihood of the ego being swallowed up
again. This whole process often takes place in the transference.
Since exaggerated symbiotic behavior is a relationship prob-

lem, it is to be expected that all forms of symbiosis reappear in therapy as problems of transference and countertransference.

The tale "Redhair Greeneyes" described issues pertaining to authority and father-boundedness, masculine identity, and difficulties in relationships with women. Here, the way out of symbiosis involved strengthening the ego by integrating what had been repressed (embodied in the figure of the companion, Redhair Greeneyes). This integration then opened up a greater capacity for relating to the feminine. Naturally, symbiosis with the companion had also to be resolved. The role of the companion as laid out here may be taken over by the analyst in a therapeutic situation, but it can also be represented by an inner, psychic figure or a friend in the outer, social world. It seems to me that this way out of symbiosis is indicated when one's identity is symbiotically bound up with a parent of the same sex.

"The Daughter of the Lemon Tree" dealt with a depression rooted in a family system characterized by overprotectedness. The tale showed the psychological attitudes spawned by such a system: an illusory view of life, grandiose ideas, intense longing for an ideal partner—but also a large capacity for trust. Within this general psychodynamic, depression represents a typical reaction to disorder—as well as to hard reality. Conversely, a mild form of mania represents a typical response to separation. The way out of this situation begins with acceptance of a "simple" life and its drudgery, which can be made more bearable by keeping the grand vision in sight. It is a way that plods through reality one step at a time so as to gradually "ground" one's grandiose fantasies. By enduring the depression that follows upon the loss of grandiose ideas—and perhaps even grieving the loss of those ideas—this existential feeling takes on dimension and depth, an intense feeling that can be discovered within oneself and realized in a relationship. This process may take place in the analytical situation when an analysand finds he is no longer able to maintain an idealized

■ Notes

1. Further methodical comments on the interpretation of folktales can be found in M. Jacoby, V. Kast, I. Riedel, *Witches, Ogres, and the Devil's Daughter: Encounters with Evil in Fairy Tales*, pp. 40 ff. M.-L. von Franz has written exemplary folktale interpretations for the Jungian school, for example, *Problems of the Feminine in Fairy Tales*.

2. "Von dem Burschen, der sich vor nichts fürchtet," from *Isländische Volksmärchen*. Parallel: "The Boy Who Left Home to Find Out About the Shivers," in *Grimms' Tales for Young and Old*, no. 4. See also "The Boy Who Knew No Fear" in *Icelandic Folktales and Legends*, ed. by J. Simpson, pp. 122 ff.

3. C. G. Jung, "On Psychic Energy," in *The Structure and Dynamics of the Psyche*, *CW* 8, par. 183.

4. W. Weischedel, *Skeptische Ethik*.

5. *Handwörterbuch des Deutschen Aberglaubens*, vol. 2, p. 1286.

6. "The Goose Girl," in *Grimms' Tales for Young and Old*, no. 89.

7. B. Schliephacke, *Märchen. Seele und Sinnbild*, p. 111.

8. M. Ninck, *Wodan und Germanischer Schicksalsglaube*, p. 93.

9. "Graumantel," in *Deutsche Volksmärchen. NeueFolge*, E. Moser-Rath, ed.

10. Bolte-Polivka, *Anmerkungen zu den Kinder- und Hausmärchen der Brüder Grimm*.

11. "The Lilting, Leaping Lark," in *Grimms' Tales for Young and Old*, no. 88.

12. For another translation of this fairy tale, see "The Nixie of the Pond" in *Grimms' Tales for Young and Old*, no. 181.

13. C. G. Jung, "The Secret of the Golden Flower," *CW* vol. 13.

14. C. G. Jung, *Psychological Types*, *CW* vol. 6, par. 757, p. 448. Quotation: as "a process of differentiation (q.v.), having for its goal the development of the individual personality."

15. M. Mahler, F. Pine and A. Bergman, *The Psychological Birth of the Human Infant: Symbiosis and Individuation*, p.44.

16. Ibid.

17. "Die Ehegatten," in *Finnische und Estnische Märchen*, A. von Loewis of Menar, ed.

18. E. Fromm, *The Heart of Man: Its Genius for Good and Evil*, pp. 37 ff.

19. E. Fromm, *The Art of Loving*, p.20.

20. G. Blanck and R. Blanck, *Ego Psychology: Theory and Practice*, part 2.

21. Teresa of Avila, *The Complete Works of Saint Teresa of Jesus*, vol. 1, "Life," chapter 24, p. 155.

22. "Die Reise in die Unterwelt zur Strudelhöhle Fafá," in *Südseemärchen*, P. Hamruch, ed.

23. Judges 13:24 ff.

24. "Rothaarig-Grünäugig," in *Kurdische Märchen*, L.-C. Wentzel, ed.

25. Cf. E. Petzold, R. Achim and A. Reindell, *Klinische Psychosomatik*, p.158 ff.

26. The Tobit story has been dated at 200 B.C., although it is probably older.

27. *Die Tochter des Zitronenbaums. Märchen aus Rhodos*, M. Klaar, ed.

28. Plato, *Timaeus* 90B-90C.

29. "Jorinde und Joringel," from KHM. See also "Jorinda and Joringel," in *Grimms' Tales for Young and Old*, no. 69.

30. For another translation of *Hymns to the Night*, see Bibliography.

■ Bibliography

Blanck, Gertrude; and Blanck, Rubin. *Ego-Psychology: Theory and Practice.* New York: Columbia University Press, 1974.

Bolte-Polikovka, *Anmerkungen zu den Kinder- und Hausmärchen der Brüder Grimm.* Hildesheim: Olms, 1963.

Fromm, Erich, *The Art of Loving.* New York and Evanston: Harper & Row, 1956.

————. *The Heart of Man: Its Genius for Good and Evil.* New York, Evanston & London: Harper & Row, 1964.

Grimms' Tales for Young and Old. Trans. by Ralph Manheim. Garden City, New York: Anchor Press/Doubleday, 1977.

Hambruch, Paul, ed. *Südseemärchen.* Düsseldorf and Cologne: Eugen Diederichs Verlag, 1979.

Handwörterbuch des Deutschen Aberglaubens. Vol. 2. Berlin and Leipzig: de Gruyter, 1930/31.

Isländische Volksmärchen. Düsseldorf and Cologne: Eugen Diederichs Verlag, 1923.

Jacoby, Jolande; Kast, Verena; and Riedel, Ingrid. *Witches, Ogres and the Devil's Daughter: Encounters with Evil in Fairy Tales.* Boston and London: Shambala, 1992.

Jung, Carl Gustav. *Collected Works* (=*CW*). Edited by Gerhard Adler et al. Bollingen Series XX. Princeton: Princeton University Press, 1954 ff. The following volumes were used in particular:
CW 6: *Psychological Types.* 2nd ed. 1971
CW 8: *The Structure and Dynamics of the Psyche.* 2nd ed. 1968.
CW 13: *Alchemical Studies.* 1968

Kast, Verena. *The Nature of Loving: Patterns of Human Relationship.* Wilmette, Illinois: Chiron Publications, 1986.

————. *A Time to Mourn: Growing through the Grief Process*. Einsiedeln: Daimon, 1988.

————. *The Creative Leap: Psychological Transformation through Crisis*. Wilmette, Illinois: Chiron Publications, 1990.

————. *Joy, Inspiration, and Hope*. College Station: Texas A & M University Press, 1991.

————. *Sisyphus: The Old Stone—A New Way*. Einsiedeln, Daimon, 1991.

————. *The Dynamics of Symbols: Fundamentals of Jungian Psychotherapy*. New York: Fromm International, 1992.

Kinder- und Hausmärchen. Zurich: Manesse, 1946.

Klaar, Marianne, ed. *Die Tochter des Zitronenbaums. Märchen aus Rhodos*. Kassel: Erich Roeth-Verlag, 1970.

Mahler, Margaret S.; Pine, Fred; and Bergman, Anni. *The Psychological Birth of the Human Infant: Symbiosis and Individuation*. New York: Basic Books, 1975.

Moser-Rath, Elfriede, ed. *Deutsche Volksmärchen. Neue Folge*. Düsseldorf and Cologne: Eugen Diederichs Verlag, 1966.

Ninck, M. *Wodan und Germanischer Schicksalsglaube*. Darmstadt: Wissenschaftliche Buchgesellschaft, 1967.

Novalis. *Hymns to the Night*. New Paltz, NY: Treacle Press, 1978.

Petzold. Ernst; and Reindell, Achim. *Klinische Psychosomatik*. Heidelberg: Quelle und Meyer, 1980.

Plato. *Timaeus*.

Schliephacke, Bruno P. *Märchen. Seele und Sinnbild*. Münster: Aschendorff, 1974.

Simpson, Jaqueline, ed. *Icelandic Folktales and Legends*. London: Batsford. 1972.

Teresa of Avila; Peers, E. Allison, ed. *The Complete Works of Saint Teresa of Jesus*. New York: Sheed and Ward, 1946.

Von Franz, Marie-Louise. *Problems of the Feminine in Fairy Tales*. Dallas: Spring Publications, 1972.

Von Loewis of Menar, August, ed. *Finnische und Estnische Märchen*. Düsseldorf and Cologne: Eugen Diederichs Verlag, 1962.

Weischedel, Wilhelm. *Skeptische Ethik*. Frankfurt: Suhrkamp, 1980.

Wentzel, Louise-Charlotte, ed. *Kurdische Märchen*. Düsseldorf and Cologne: Eugen Diederichs Verlag, 1978.